Studies in Computational Intelligence

Volume 930

Series Editor

Janusz Kacprzyk, Polish Academy of Sciences, Warsaw, Poland

The series "Studies in Computational Intelligence" (SCI) publishes new developments and advances in the various areas of computational intelligence—quickly and with a high quality. The intent is to cover the theory, applications, and design methods of computational intelligence, as embedded in the fields of engineering, computer science, physics and life sciences, as well as the methodologies behind them. The series contains monographs, lecture notes and edited volumes in computational intelligence spanning the areas of neural networks, connectionist systems, genetic algorithms, evolutionary computation, artificial intelligence, cellular automata, self-organizing systems, soft computing, fuzzy systems, and hybrid intelligent systems. Of particular value to both the contributors and the readership are the short publication timeframe and the world-wide distribution, which enable both wide and rapid dissemination of research output.

Indexed by SCOPUS, DBLP, WTI Frankfurt eG, zbMATH, SCImago.

All books published in the series are submitted for consideration in Web of Science.

More information about this series at http://www.springer.com/series/7092

Haengkon Kim · Roger Lee
Editors

Software Engineering in IoT, Big Data, Cloud and Mobile Computing

 Springer

Editors
Haengkon Kim
School of IT Engineering
Daegu Catholic University
Daegu, Korea (Republic of)

Roger Lee
ACIS International
Mount Pleasant, MI, USA

ISSN 1860-949X ISSN 1860-9503 (electronic)
Studies in Computational Intelligence
ISBN 978-3-030-64775-9 ISBN 978-3-030-64773-5 (eBook)
https://doi.org/10.1007/978-3-030-64773-5

This Springer imprint is published by the registered company Springer Nature Switzerland AG
The registered company address is: Gewerbestrasse 11, 6330 Cham, Switzerland

Preface

The purpose of International Semi-Virtual Workshop on Software Engineering in IoT, Big Data, Cloud and Mobile Computing (SE-ICBM 2020) held on October 17, 2020, Soongsil University, Seoul, Korea, is aimed at bringing together researchers and scientists, businessmen and entrepreneurs, teachers and students to discuss the numerous fields of computer science and to share ideas and information in a meaningful way. This special session Software Engineering with ICBM will discuss the analysis, evaluation and implementation of IoT, big data, cloud and mobile computing as well as the appreciation of mobile platform development. Through the special session, advanced hands-on theory and practical projects approach.

This publication captures 17 of the conference's most promising papers. The selected papers have not been published in the workshop proceedings or elsewhere, but only in this book. We impatiently await the important contributions that we know these authors will bring to the field.

In Chapter "Variability Modeling in Software Product Line: A Systematic Literature Review," Aman Jaffari, Jihyun Lee and Eunmi Kim introduce the majority of the studies proposed techniques for modeling variability in a separate model rather than modeling variability as an integral part of the development artifact.

In Chapter "The Use of Big Data Analytics to Improve the Supply Chain Performance in Logistics Industry," Lai Yin Xiang, Ha Jin Hwang, Haeng Kon Kim, Monowar Mahmood and Norazryana Mat Dawi discuss a bigger picture of how the use of big data analytics can improve the supply chain performance in logistics industry. Logistics industry could benefit from the results of this research by understanding the key success factors of big data analytics to improve supply chain performance in logistics industry.

In Chapter "Development of Smart U-Health Care Systems," Symphorien Karl Yoki Donzia and Haeng Kon Kim propose a personalized service to each user to provide differentiated medical care and change their perception of how to provide services smart U-healthcare. The achievement of service-oriented smart healthcare information systems focused on the healthcare cloud environment.

In Chapter "An Implementation of a System for Video Translation Using OCR," Sun-Myung Hwang and Hee-Gyun Yeom develop an image translator that combines existing OCR technology and translation technology and verify its effectiveness. Before developing, we presented what functions are needed to implement this system and how to implement them, and then tested their performance.

In Chapter "Research on Predicting Ignition Factors Through Big Data Analysis of Fire Data," Jun-hee Choi and Hyun-Sug Cho present that an artificial neural network algorithm was applied to infer the ignition factors using about 10 data per fire accident, and the prediction accuracy was about 80%, rather than determining the ignition factors.

In Chapter "Development of U-Health Care Systems Using Big Data," Symphorien Karl Yoki Donzia, Yong Pil Geum and Haeng Kon Kim propose an algorithm to detect driver drowsiness through analysis of heart rate variability and compare it with EEG-based sleep scores to verify the proposed method in bigdata. The ECG sensor provides various detection methods to detect RR interval big data from ECG data and only transmit abnormal data.

In Chapter "Analyses on Psychological Steady State Variations Caused by Auditory Stimulation," Jeong-Hoon Shin presented a technique for applying the stimuli allowing all subjects to maintain their psychologically steady state by using the customized stimuli optimized for individual users and has attempted to promote the psychologically steady state of the subjects by using stimuli tailored to each user.

In Chapter "Flipped Learning and Unplugged Activities for Data Structure and Algorithm Class," Jeong Ah Kim present that unplugged activities are designed to evaluate how much the students studied in pre-class and identify the part which students cannot understands. For post-class, just homework for identifying the application of algorithm in real life is issued.

In Chapter "Study on Agent Architecture for Context-Aware Services," Symphorien Karl Yoki Donzia and Haeng-Kon Kim proposed the general process of a situational contexed awareness system and the design considerations for the situational awareness system. In addition, it summarizes the characteristics of the existing situational awareness system and presents a trend analysis of the system. In conclusion, we propose future work for a better situational awareness system.

In Chapter "Secure Transactions Management Using Blockchain as a Service Software for the Internet of Things," Prince Waqas Khan and Yung-Cheol Byun describe suspicious transactions in IoT systems and manage them using the blockchain as a service software plans. This study builds software-specific components for blockchain functions to implement in IoT networks.

In Chapter "Plant Growth Measurement System Using Image Processing," Meonghun Lee, Haeng Kon Kim and Hyun Yoe design a smart barn monitoring system based on LoRa, a low-power long-range wireless communication technology, to supplement the wide communication range and safety when operating large-scale smart cattle shed.

In Chapter "Smart Cattle Shed Monitoring System in LoRa Network," Meonghun Lee, Haeng Kon Kim and Hyun Yoe describe the design of "smart cattle shed" monitoring system based on LoRa, a low-power, long-range wireless communication technology, to support the range of communications and safety measurements required when operating large-scale smart cattle shed. The proposed system wirelessly collects real-time stable information from sensors installed in the cattle shed, and the collected data are analyzed by the integrated management system, delivered to the user and controlled by the application.

In Chapter "The Triple Layered Business Model Canvas for Sustainability in Mobile Messenger Service," Hyun Young Kwak, Myung Hwa Kim, Sung Taek Lee and Gwang Yong Gim proposed that digital technology has a positive effect not only on economic performance but also on environmental and social aspects. By applying it to the service, we were able to confirm that digital technological innovation can achieve corporate sustainability.

In Chapter "The Effects of Product's Visual Preview on Customer Attention and Sales Using Convolution Neural Networks," Eun-tack Im, Huy Tung Phuong, Myung-suk Oh, Simon Gim and Jun-yeob Lee, regression analyzed how images provided by the sellers affect product sales page views and product sales. Using attributes such as emotion, aesthetics and product information were extracted through deep-CNNs model and vision API from t-shirts images sold in Singapore's leading mobile commerce application "Shopee." As a result of the analysis, the image information entropy and the color-harmony representing the emotion of the image had a significant effect on the number of views.

In Chapter "Implementation of Cloud Monitoring System Based on Open Source Monitoring Solution," Eunjung Jo and Hyunjae Yoo proposed that a cloud monitoring system based on an open source was designed and implemented by subdividing it into CPU, memory, storage and network parts. Linux and KVM operating systems were used to implement the designed system, and Zabbix and Grafana were used as open-source monitoring solutions.

In Chapter "A New Normal of Lifelong Education According to the Artificial Intelligence and EduTech Industry Trends and the Spread of the Untact Trend," Cheong-Jae Lee and Seong-Woo Choi present a new normal of lifelong education as a lifelong education classification system innovation, self-directed learning ability cultivation, future-oriented entrepreneurship and startup education, collaborative literacy education, maker's experience, adaptive learning system and digital literacy education platform.

In Chapter "The Optimal Use of Public Cloud Service Provider When Transforming Microservice Architecture," Sungwoong Seo, Myung Hwa Kim, Hyun Young Kwak and Gwang Yong Gim provide an optimal environment even in situations where many reviews of hybrid cloud for backup, recovery and expansion purposes of enterprises (utilizing infrastructure transition between on-premises and public cloud).

It is our sincere hope that this volume provides stimulation and inspiration, and that it will be used as a foundation for works to come.

Daegu, Korea   Haengkon Kim
August 2020

Contents

Contributors

Yung-Cheol Byun Department of Computer Engineering, Jeju National University, Jeju-si, Republic of Korea

Hyun-Sug Cho Department of Fire and Disaster Prevention, Daejeon University, Daejeon, South Korea

Jun-hee Choi Department of Disaster Prevention, Graduate School, Daejeon University, Daejeon, South Korea

Seong-Woo Choi Department of Lifelong Education, Soongsil University, Seoul, Korea

Norazryana Mat Dawi Sunway University, Subang Jaya, Malaysia

Symphorien Karl Yoki Donzia Department of Computer Software, Daegu Catholic University, Gyeongsan-si, South Korea

Yong Pil Geum Department of Innovation Start-Up and Growth, Daegu Catholic University, Gyeongsan-si, South Korea

Gwang Yong Gim Department of Business Administration, Soongsil University, Seoul, South Korea

Simon Gim SNS Marketing Research Institute, Soongsil University, Seoul, South Korea

Ha Jin Hwang Sunway University, Subang Jaya, Malaysia

Sun-Myung Hwang Daejeon University, Daejeon, South Korea

Eun-tack Im Graduate School of Business Administration, Soongsil University, Seoul, South Korea

Aman Jaffari Department of Software Engineering, Jeonbuk National University, Jeonju, Republic of Korea

Eunjung Jo Soongsil University, Seoul, Korea

Prince Waqas Khan Department of Computer Engineering, Jeju National University, Jeju-si, Republic of Korea

Eunmi Kim Department of Computer & Game Convergence, Howon University, Gunsan, Republic of Korea

Haeng-Kon Kim Department of Computer Software, Daegu Catholic University, Gyeongsan-si, South Korea

Haeng Kon Kim School of Information Technology, Catholic University of Daegu, Gyeongsan, Gyeongbuk-do, Republic of Korea;
Daegu Catholic University, Gyeongsan, South Korea

Jeong Ah Kim Department of Computer Education, Catholic Kwandong University, Gangneung, South Korea

Myung Hwa Kim Department of IT Policy and Management, Soongsil University, Seoul, South Korea;
Graduate School of IT Policy and Management, Soongsil University, Seoul, South Korea

Hyun Young Kwak Department of IT Policy and Management, Soongsil University, Seoul, South Korea;
Graduate School of IT Policy and Management, Soongsil University, Seoul, South Korea;
Department of Business Administration, Soongsil University, Seoul, South Korea

Cheong-Jae Lee Department of Lifelong Education, Soongsil University, Seoul, Korea

Jihyun Lee Department of Software Engineering, Jeonbuk National University, Jeonju, Republic of Korea

Jun-yeob Lee College of Economics, Sungkyunkwan University, Seoul, South Korea

Meonghun Lee Department of Agricultural Engineering, National Institute of Agricultural Sciences, Wanju-gunJeollabuk, Jeollabuk-do, Republic of Korea

Sung Taek Lee Department of Computer Science, Yong In University, Yongin, South Korea

Monowar Mahmood KIMEP University, Almaty, Kazakhstan

Myung-suk Oh Graduate School of Business Administration, Soongsil University, Seoul, South Korea

Huy Tung Phuong Graduate School of Business Administration, Soongsil University, Seoul, South Korea

Sungwoong Seo Department of IT Policy and Management, Soongsil University, Seoul, South Korea

Jeong-Hoon Shin Daegu Catholic University, Gyeongsan-si, South Korea

Lai Yin Xiang Sunway University, Subang Jaya, Malaysia

Hee-Gyun Yeom Daejeon University, Daejeon, South Korea

Hyun Yoe Department of Information and Communication Engineering, Sunchon National University, Suncheon, Jeollanam-do, Republic of Korea

Hyunjae Yoo Soongsil University, Seoul, Korea

Variability Modeling in Software Product Line: A Systematic Literature Review

Aman Jaffari, Jihyun Lee, and Eunmi Kim

Abstract Variability is the core concept characterizing software product line engineering. Over the past decades, variability modeling has been an emerging topic of extensive research that resulted in different units of variability (e.g., feature, decision, orthogonal, UML) with various variability modeling techniques. Hence, there is a need for a comprehensive study to shed light on the current status, diversity, and direction of the existing variability modeling techniques. The main objective of this research is to characterize the diversity of modeling variability and provide an overview of the status of existing literature. We conducted a systematic review with six formulated research questions and evaluated 74 studies published between the years 2004–2007. The results indicated that the majority of the studies proposed techniques for modeling variability in a separate model rather than modeling variability as an integral part of the development artifact, and the feature model was found as the most common unit of variability. Our study also identified more ambiguity in handling complexity issues as well as the need for a commonly accepted way of addressing variability model evolution. The strength of the evidence in support of the proposed approaches with illustrative examples and lack of robust tooling support that have confined the generalizability of the existing studies need further improvement with more robust empirical studies.

Keywords SPLE · Variability modeling · Feature model · Decision model · Orthogonal variability model

A. Jaffari · J. Lee (✉)
Department of Software Engineering, Jeonbuk National University, Jeonju, Republic of Korea
e-mail: jihyun30@jbnu.ac.kr

A. Jaffari
e-mail: aman@jbnu.ac.kr

E. Kim
Department of Computer & Game Convergence, Howon University, Gunsan, Republic of Korea
e-mail: ekim@howon.ac.kr

1

1 Introduction

Product line variability that is documented in so-called variability models is one of the key properties characterizing software product line engineering [1, 2]. The variability models are the most well-known promising approach for representing the commonalities and variabilities in variable systems and software product lines (SPL). Variability can be modeled either as an integral part of the development artifacts or in a separate variability model. There are alternative units of variability in the literature for representing variability models such as the feature model, decision model, and orthogonal variability model (OVM), UML, etc. Modeling variability in SPL is a very complex task that requires a good decision on how to select the most appropriate techniques and tools. Also, because software products evolve constantly, handling the changes in the variability model and its evolution is an even more challenging task [3].

Over the years, to address the variability modeling challenges, numerous variability modeling approaches have been introduced. Hence, the SPL practitioners and researchers are always overwhelmed with the massive amount of information. There is a need to shed light on the current status and direction of the existing research techniques and provide a context to properly position new research directions by summarizing the growing bodies of research for modeling variability in SPL. This research extends a conference paper we presented in KCSE 2018[1] intending to provide a comprehensive comparison of different dialects of feature modeling techniques and summarizing the growing bodies of research that can also provide a context for positioning new research directions for variability modeling in SPL.

The results indicated that the majority of the studies favored separated variability modeling instated of modeling variability as an integral part of the development artifact following a single feature modeling technique rather than multiple feature modeling with feature models as the units of variability. Even though most of the studies do not properly specify handling the variability modeling complexity, the hierarchical model decomposition was adopted as a more common approach to handle complexity. Besides, the majority of studies do not handle the variability model evolution, and there is a lack of real-world industrial case studies and tools.

The rest of the paper is organized as follows. Section 2 presents the background of the study and discusses the related works. Section 3 describes the method for our systematic review study. The result of the study is presented in Sect. 4. A brief discussion of our findings is presented in Sect. 5. And finally, Sect. 6 concludes the study and presents future work.

---

[1]Jaffari, A., Lee, J., Jo, J. H., Yoo, C. J.: A Systematic Review on Variability Modeling in Software Product Line. In proceedings of the 20 Korean Conference on Software Engineering (KCSE 2018).

2 Background and Related Work

As the importance of documenting and managing variability increase in SPL, the research communities are continuously in the quest of finding a better solution for modeling variability. Over the last decade, researchers have been proposed a significant number of techniques and tools for modeling variability. Some of which are the extension or modification of the existing works while many others introduce or develop their concepts. An explanation of such an issue is given by [4] referring to different "dialects" of feature modeling techniques. As a result, SPL practitioners are overloaded with the growing, diverse terminologies and concepts so that even the most experienced SPL practitioner can get easily lost. Categorization and summarization of all these concepts became vital more than any time. A study on software product line engineering and variability management [1] has briefly summarized highlights of the research achievements in SPLE focused on overall concepts, techniques, and methods. The styles in variability modeling and the appropriateness of different approaches for modeling process variability are investigated by [5] and [6].

A systematic mapping study conducted by [7] to analyze the important aspects that should be considered when adopting SPL approaches. How to describe SPL variabilities in textual use cases is indicated by [8]. An investigation on the use of visualization techniques to SPL development by [9] revealed the feature models as a de facto standard for describing the combinations of features in the products of an SPL. The two most relevant papers to this study are [10], an investigation on classifying variability modeling techniques, and [11] a survey on industrial variability modeling techniques. The first one provides an overview of the variability modeling techniques using a single example to exemplify the concepts of all techniques. The study focusses on identifying the common characteristics of variability modeling techniques. The second one, emphasis on understanding the use of variability modeling in practice and the associated benefits. However, besides using a different approach, these studies are missing the recent trends and dialects of modeling variability in SPL.

There are comparable studies to our study for identifying the challenges and classifying variability modeling techniques. However, these comparable studies date back to an average of more than six to seven years. And so, they are missing most of the recent variability modeling concepts and trends in the research area.

3 Systematic Mapping Method

A systematic literature review is an exhaustive and systematic method for identifying, evaluating, and synthesizing the best quality scientific studies related to a specific topic area or research question. The significance of systematic literature reviews and systematic mapping studies that complement systematic reviews for software engineering has been discussed by many studies [12–14]. There are many guidelines for performing systematic reviews. We performed similar steps to guidelines provided by [14, 15] for systematic literature review in software engineering.

To specify our primary goal, we have defined the following research questions:

- **RQ1**: Which variability modeling approaches are proposed in SPL?
- **RQ2**: How do the variability modeling approaches in SPL handle the complexity issues?
- **RQ3**: How do the variability modeling approaches in SPL handle the variability model evolution?
- **RQ4**: What is the strength of the evidence in support of the proposed variability modeling approaches?
- **RQ5**: Do the variability modeling approaches in SPL support tools?
- **RQ6**: What are the implications of these findings for the SPL research community?

Study selection process: Our study selection process is divided into six phases with the corresponding activities (Fig. 1). The "apply search in databases" activity in Phase 1 returned 1192 studies. After refining search strings in Phase 2, a total of 976 studies were excluded by re-performing the search in the databases, filtering duplicated studies, and merging relevant studies. As a result, 216 studies were selected. In Phase 3, we excluded 115 more studies by reading the title, abstracts, and keywords, applying inclusion and exclusion criteria. In Phase 4, we performed complete text reading and as a result, 76 studies were selected. By performing the snowball sampling activity in Phase 5, additional 6 studies were included mostly related to variability management. Finally, by quality assessment activity in Phase 6, a total of 8 more studies were excluded that showed insignificant contribution.

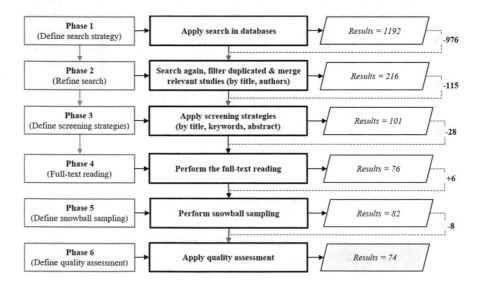

Fig. 1 An overview of the study selection process

Table 1 Search string and associated results in databases

Databases	URL	Search string	Results	
			Raw	Refined
IEEE	https://www.ieee.org/	(variability model OR feature model OR decision model) AND (product lines)	553	113
ACM	https://dl.acm.org/	("variability model" "feature model" "decision model")	362	84
ScienceDirect	https://www.sciencedirect.com/	(("variability model" or "feature model" or "decision model") and ("product line" or "product lines" or "product family"))	194	16
Springer	https://link.springer.com/	"variability model" "feature model" "decision model"	83	3

Search in databases: The search strategy is performed to extract the most relevant studies. Table 1 lists the Databases, URL, Search String, and associated results.

Screening Strategies: Prior to applying screening strategies for the study selection, we make sure that there are no duplicated studies. We also merge relevant studies with the same authors on the same topic. To further reduce the likelihood of bias and to assure that the most relevant papers in the scope of modeling variability in SPL are detected, we have defined a number of inclusion criteria (IC) and exclusion criteria (EC) as follows:

- *IC-1*: Studies that explore and provides new techniques and tools, advance the existing techniques and tools for modeling variability in SPL. *IC-2*: Studies that explore and provide solutions for handling complexity issues for modeling variability in SPL. *IC-3*: Studies that present techniques and solutions for handling variability model evolution in SPL.
- *EC-1*: Studies that lie outside the scope of variability modeling in SPL (since some variability modeling related terms and concepts are used in other areas of study). *EC-2*: Short papers, review papers, and papers that compare and contrast two or more different variability modeling techniques in SPL. *EC-3*: Studies that provide no evidence in support of the variability modeling approach.

Snowball sampling: The snowball sampling in systematic literature reviews for identifying additional studies is very common using the reference list of studies or the citations to the studies. We performed our snowball sampling similar to a guideline provided by [16]. We focused more on snowballing the studies that used the term variability management as an alternative to variability modeling.

Quality assessment: The quality assessment on each of the 82 studies, was performed answering the following questions that are initially provided by [17, 18].

- Is the selected paper based on research and it is not merely a "lessons learned" report based on an expert opinion? Is there a clear statement of the aims of the research? Is there an adequate description of the context in which the research was carried out? Was the research design appropriate to address the aims of the research? Was the data analysis sufficiently rigorous? Is there a clear statement of findings?

4   The Results

In this section, we will answer to each of the defined RQs based on the extracted information during the review process.

The proposed approaches (RQ1): As depicted in Fig. 2 left, about 74% of the studies followed the variability modeling techniques in a separate variability model, and only 21% of the studies performed modeling variability as an integral part of development artifacts. Also, as shown in Fig. 2 right, 60% of the studies preferred feature models and 31% favored variation points, whereas very few studies have adopted decision modeling as the units of variability. The remaining 4% introduced their own concepts and terminology, for example, "clafer" that is a unifying concept for expressing feature models more concisely [19]. As another example, we can point out to some UML based feature modeling [20].

Handling complexity (RQ2): As depicted in Fig. 3 right, most of the researchers favored single feature modeling techniques over multiple feature modeling that is 66% against 24%, and we were not able to distinguish the remaining 10% in terms of the model multiplicity.

Also, we investigated handling the complexity of the variability model from the three perspectives: hierarchical model decomposition, modularity, and viewpoints. When compared to modularity and viewpoints, the hierarchical model decomposition is adopted as a more common means for handling complexity issues. However, a large portion of the studies never stated exactly how do they handled the complexity and by what means.

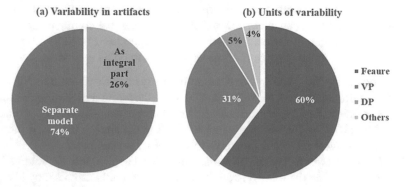

Fig. 2 Distribution of the studies by adopting variability in artifacts and units of variability

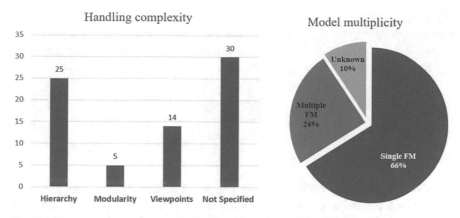

Fig. 3 Distribution of studies by handling complexity and model multiplicity

Variability model evolution (RQ3): We have investigated handling the variability model evolution from two perspectives, consistency check, and conflict resolution. As shown in Fig. 4 right, the majority of the studies do not handle the variability model evolution. A handful of the most remarkable solutions we found in the literature are: first, handling variability model evolution together with related artifacts, by inspecting a sample of the evolution history [21]. The main idea is to analyze the usage frequency of each pattern and how artifacts are affected when there is a change in the model. Second, handling the variability model evolution by enabling model composition with interfaces [22]. Via composition, a variability model is allowed to be included as an instance to another feature model. Similar to programming interfaces, a variability model interface is used for the composition of variability models to establish information hiding. Third, handling the variability model evolution by locking specific configurations [23]. The idea is to add temporal variability (feature) models to assess the impact on existing configurations.

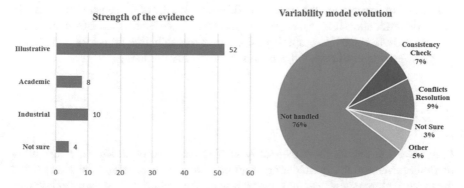

Fig. 4 Distribution of papers by the strength of the evidence and variability model evolution

Fig. 5 Distribution of
papers by tooling support

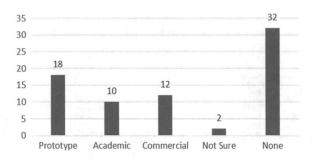

The strength of the evidence (RQ4): Fig. 4 left indicates that all the researches
performed a preliminary assessment, either using illustrative examples, academic
studies, or industrial practices as the strength of the evidence in support of their
proposed approaches. However, most of them are using illustrative examples,
whereas, a few studies conducted academic, and a few others claimed using industrial
examples.

Tool support (RQ5): Tooling support is crucial for modeling variability in
SPL, particularly for handling complexity and facilitating model evolution in large
complex models. However, as depicted in Fig. 5, most of the studies in the
current literature did not provide a realistic tool for variability model formation
and manipulation assessment.

The findings' implications (RQ6): The answers we have revealed so far via the
preceding research questions can be used to determine the possible implications of
the findings. For instance, the trends in high adoption of single feature model and
feature as units of variability (RQ1). More ambiguity on how the complex issues
are handled for modeling variability in SPL (RQ2). Lack of a commonly accepted
way of addressing variability model evolution (RQ4). Presenting a large number
of non-realistic examples as a means for the strength of the evidence in support
of the proposed methods (RQ4). Lack of robust tooling support by adopting more
illustrative tools (RQ5). The practical implication of this study is providing well-
organized information on the status and different dialects of the existing literature in
variability modeling from a different perspective. Table 2 provides information that
can be used as a context to properly position new research directions by summarizing
the growing bodies of research concerning variability modeling.

5 Discussion

In general, the results show that many studies have preferred separate variability
modeling over variability modeling as an integral part of development artifact, and
the single feature model over multiple feature models. Multiple feature models can be
used to reduce the model complexity by dividing the model into smaller manageable
models. However, there is always a tradeoff between the model complexity and

Table 2 A comprehensive comparison of selected studies

Selected studies	RQ1					RQ2					RQ3		RQ4		RQ5			
	IPD	SVM	F	DP	VP	H	M	V	SFM	MFM	CC	CR	IE	AS	IP	IT	AT	CIT
P01	✓	•	✓	•	•	✓	•	•	•	Ø	✓	•	✓	•	•	•	•	•
P02	•	✓	✓	•	•	✓	•	•	Ø	•	•	✓	•	•	✓	•	•	✓
P03	•	✓	✓	•	•	✓	•	•	✓	•	✓	•	•	•	✓	✓	•	•
P04	•	✓	•	•	✓	Ø	•	•	✓	•	•	✓	✓	•	•	•	•	•
P05	•	✓	•	•	✓	✓	•	•	✓	•	•	•	✓	•	•	•	Ø	•
P06	✓	•	✓	•	•	✓	•	•	✓	•	•	•	•	•	✓	✓	•	✓
P07	•	✓	✓	•	✓	•	•	✓	•	✓	•	✓	✓	•	•	•	✓	•
P08	•	✓	•	•	•	•	•	✓	•	✓	Ø	•	•	✓	•	✓	•	•
P09	•	•	✓	•	✓	•	•	•	✓	•	•	•	•	•	Ø	•	•	•
P10	✓	✓	✓	•	•	•	•	✓	✓	✓	•	✓	✓	•	✓	✓	✓	✓
P11	•	•	✓	•	•	•	•	•	•	•	•	•	✓	•	•	•	•	•
P12	•	✓	✓	•	•	✓	•	•	✓	✓	•	•	✓	•	•	•	✓	•
P13	•	✓	✓	•	•	✓	•	•	•	•	•	•	✓	•	•	✓	•	•
P14	•	✓	•	•	•	✓	✓	•	✓	✓	Ø	•	✓	•	•	•	•	✓
P15	•	✓	✓	•	✓	✓	•	•	✓	•	✓	•	✓	✓	•	•	✓	•
P16	✓	•	✓	•	•	✓	•	•	•	✓	•	•	•	•	Ø	•	•	•
P17	✓	•	✓	•	•	•	•	✓	✓	•	✓	✓	✓	✓	•	•	✓	✓
P18	•	✓	✓	•	•	•	•	✓	•	✓	•	•	✓	•	•	•	•	•
P19	•	✓	•	•	✓	•	•	✓	•	✓	•	•	✓	•	•	•	•	•
P20	✓	•	•	•	✓	✓	•	•	•	✓	•	•	✓	•	•	•	•	✓

(continued)

Table 2 (continued)

Selected studies	RQ1					RQ2					RQ3		RQ4			RQ5		
	IPD	SVM	F	DP	VP	H	M	V	SFM	MFM	CC	CR	IE	AS	IP	IT	AT	CIT
P21	•	✓	✓	•	•	•	•	✓	✓	•	•	•	✓	∅	•	•	•	•
P22	✓	•	✓	•	•	•	•	•	✓	•	•	•	•	•	•	✓	•	•
P23	✓	•	✓	•	•	•	•	•	✓	•	•	•	✓	✓	•	•	•	✓
P24	•	✓	✓	•	•	✓	•	•	•	✓	•	•	•	•	∅	•	•	∅
P25	✓	•	•	•	✓	•	•	•	✓	•	•	•	✓	•	•	•	•	•
P26	•	✓	✓	•	•	✓	•	•	✓	•	•	•	✓	•	•	•	•	•
P27	•	✓	✓	•	•	✓	•	•	✓	•	•	•	✓	•	•	•	•	•
P28	•	•	•	•	✓	•	✓	•	•	•	•	•	✓	•	•	•	✓	•
P29	✓	✓	✓	•	•	•	•	✓	•	✓	•	•	✓	•	•	✓	•	•
P30	•	✓	•	•	✓	•	•	✓	✓	•	•	•	✓	•	•	•	•	•
P31	•	✓	✓	•	•	•	•	•	✓	•	•	•	✓	•	•	•	•	•
P32	•	✓	✓	•	•	✓	•	•	✓	•	•	•	✓	•	•	•	•	•
P33	•	•	•	•	✓	✓	•	✓	✓	•	•	•	✓	•	•	•	•	•
P34	•	✓	✓	•	•	•	•	•	✓	•	•	•	✓	•	•	•	•	•
P35	✓	✓	✓	•	•	✓	•	•	•	✓	•	•	•	✓	•	•	✓	•
P36	•	•	•	•	✓	✓	•	•	•	✓	•	•	•	✓	•	•	✓	•
P37	•	✓	•	•	✓	✓	•	•	✓	•	✓	•	✓	•	•	•	•	•
P38[21]	✓	•	•	•	✓	✓	•	•	∅	•	•	•	•	•	✓	•	•	•
P39	•	✓	✓	•	•	✓	•	•	✓	•	•	•	✓	•	•	✓	•	•

(continued)

Table 2 (continued)

Selected studies	RQ1		RQ2								RQ3		RQ4			RQ5		
	IPD	SVM	F	DP	VP	H	M	V	SFM	MFM	CC	CR	IE	AS	IP	IT	AT	CIT
P40	✓	•	•	•	✓	•	✓	•	∅	•	•	•	✓	•	•	•	•	•
P41	•	✓	•	•	✓	•	•	✓	•	✓	•	•	✓	•	•	•	•	•
P42	•	✓	✓	•	•	•	•	✓	•	✓	•	✓	•	•	✓	•	✓	•
P43	•	✓	○	○	○	✓	•	•	∅	∅	•	•	✓	•	•	•	•	•
P44	•	✓	○	○	○	✓	•	•	✓	•	•	•	✓	•	•	✓	•	•
P45	•	✓	•	•	✓	•	•	•	✓	•	•	•	✓	•	•	•	•	•
P46	•	✓	✓	•	•	•	•	•	✓	•	•	•	•	✓	•	✓	•	•
P47	•	✓	✓	•	•	•	•	•	✓	•	•	✓	✓	•	•	✓	•	•
P48	•	✓	✓	•	•	•	•	•	✓	•	•	•	✓	•	•	✓	•	•
P49	•	✓	•	✓	•	•	•	•	✓	•	•	•	✓	•	•	•	•	•
P50	•	✓	✓	•	•	•	•	•	∅	•	•	•	✓	•	•	✓	•	•
P51	•	✓	✓	•	•	•	•	•	✓	•	•	•	✓	•	•	✓	•	•
P52	•	✓	✓	•	•	•	•	•	✓	•	•	•	✓	•	•	•	•	•
P53	•	✓	✓	•	•	•	•	•	✓	•	•	•	✓	•	•	•	•	•
P54	•	✓	•	•	✓	•	•	✓	✓	•	•	•	✓	•	•	✓	✓	•
P55	•	✓	✓	•	✓	✓	•	•	•	•	•	✓	✓	•	•	•	•	•
P56	•	✓	○	○	○	○	○	○	.	•	•	•	✓	•	•	•	•	•
P57	•	✓	✓	•	•	•	•	•	.	.	•	•	✓	•	•	•	•	•
P58	•	✓	✓	•	•	•	•	•	✓	•	•	•	✓	•	•	✓	•	•

(continued)

Table 2 (continued)

Selected studies	RQ1				RQ2						RQ3		RQ4			RQ5		
	IPD	SVM	F	DP	VP	H	M	V	SFM	MFM	CC	CR	IE	AS	IP	IT	AT	CIT
P59	✓	•	✓	•	•	•	•	•	✓	•	•	•	•	✓	•	•	•	•
P60	•	✓	✓	•	•	•	•	•	✓	•	•	•	✓	•	•	•	•	•
P61	•	✓	✓	•	•	•	✓	•	•	✓	•	•	✓	•	•	✓	•	•
P62	•	✓	✓	•	•	•	•	•	✓	•	•	•	✓	•	•	✓	•	•
P63	✓	✓	✓	•	•	•	•	•	✓	•	•	•	•	•	✓	•	•	✓
P64	•	✓	✓	•	•	•	•	•	✓	•	•	•	•	•	✓	•	✓	•
P65	•	✓	✓	•	✓	•	•	•	✓	•	•	•	✓	•	•	•	•	✓
P66	•	✓	•	•	✓	•	•	•	✓	•	O	O	•	•	✓	•	•	✓
P67	✓	•	•	•	✓	•	•	•	✓	•	•	•	✓	•	•	•	•	•
P68	•	✓	✓	•	•	•	•	•	✓	•	•	•	✓	•	•	•	•	•
P69	•	✓	✓	•	•	•	•	•	✓	•	•	•	✓	•	•	•	•	✓
P70	✓	•	•	•	✓	•	•	✓	•	✓	•	•	•	•	✓	•	•	✓
P71	✓	•	•	•	✓	•	•	✓	•	✓	•	•	•	•	•	•	•	✓
P72	✓	•	•	•	✓	•	•	•	✓	•	•	•	•	•	•	✓	•	•
P73	•	✓	•	•	✓	✓	•	•	✓	•	•	•	✓	•	•	•	•	•
P74	•	✓	✓	•	•	•	•	•	•	✓	•	•	✓	•	•	•	•	✓

Legend IPD: As Integral Parts of Development Artifacts, *SVM*: Separate Variability Model, *F*: Feature, *DP*: Decision Point, *VP*: Variation Point, *H*: Hierarchy, *M*: Modularity, *V*: View Point, *SFM*: Single Feature Model, *MFM*: Multiple Feature Model, *CC*: Consistency Check, *CR*: Conflicts Resolution, *IE*: Illustrative Example, *AS*: Academic Study, *IP*: Industrial Practice, *IT*: Illustrative Tool (Prototype or Small Tool), *AT*: Academic Tool, *CIT*: Commercial Industry Practice, [✓]: Characteristic correspond to the question, [•]: Characteristic do not correspond to the question, [Ø]: Not Sure, O: Others

Note Due to page limitations, the given references for selected studies are available online at the following GitHub page: https://amanjafari.github.io/VMSPL/

consistency. In the case of multiple feature models, the variability is distributed over multiple models. Hence it is hard to keep the consistency unless there is sufficient tooling support. On the other hand, placing all the variability in a single feature model can lead to greater complexity.

Lack of robust tooling support for managing consistency can be a good reason for adopting a single feature model to avoid the variability dispersions across different models. Also, the single feature modeling has the maintainability and understandability problems because including every variability in a single model make the model grow larger and more complex. Recently, BigLever software published its vision about feature models for mega-scale product line. BigLever software proposes a hierarchical feature modeling concepts composed of primitive standalone feature, feature model as a bundle of primitive standalone features, subsystem product lines as a bundle of feature models, and system product lines by bundling subsystem product lines.

Until now, the modeling variability in a single feature model is preferred by many researchers due to the easiness of validating their proposed approaches, especially with very simple illustrative examples. However, modeling an extremely large real-world SPL in a single feature model seems not to be realistic and even impossible. Such problems are partially addressed by researchers with introducing techniques like hierarchical variability modeling (e.g., BigLever) and multiple feature modeling. Yet, problems such as interfacing between low-level abstraction and higher-level abstraction, consistency check, and model evolution further need to be addressed. Specifically, based on this systematic review, we summarize the following issues: Lack of robust tooling support; Lack of a comprehensive and commonly accepted mechanism for handling variability model evolution; More ambiguity on handling the variability modeling complexity issues, and lack of real-word industrial examples.

6 Conclusion and Future Work

In this research, we have performed a systematic literature review of variability modeling in SPL with the purpose to provide a comparison of different dialects of feature modeling techniques. We have investigated 74 studies out of 976 published between the year 2004–2017. We used different screening strategies to obtain the most relevant studies. The findings indicate that there is a major shift in the research direction from modeling variability as an integral part of development artifact to modeling variability in a separate model, and feature model is the most commonly adopted technique for units of variability. The result also indicates that there is a lack of real-world industrial tools and cases, which confines the generalizability of the proposed approaches and tools. Based on the aforementioned issues we have discussed earlier several opportunities for the future research direction are highlighted. As part of our future work, we will further investigate on tooling support and finding solutions for solving variability modeling complexity issues as well as looking for an appropriate mechanism to handle variability model evolution.

Acknowledgements This research was supported by Basic Science Research Program through the National Research Foundation of Korea (NRF) funded by the Ministry of Education (NRF-2017R1D1A3B03028609) and supported by the National Research Foundation of Korea (NRF) grant funded by the Korea government (MSIT) (NRF-2020R1F1A1071650.

References

1. Metzger A, Pohl K (2014) Software product line engineering and variability management: achievements and challenges. In: Proceedings of the on future of software engineering, Hyderabad, India, pp 70–84
2. Pohl K, Böckle G, van Der Linden FJ (2005) Software product line engineering: foundations, principles and techniques. Springer Science & Business Media
3. Kang S et al (2017) Critical analysis of feature model evolution. In: 2017 8th IEEE international conference on software engineering and service science, pp 62–65
4. Zaid LA, Kleinermann F, Troyer OD (2010) Feature assembly: a new feature modeling technique. In: Proceedings of the 29th international conference on conceptual modeling, Vancouver, BC, Canada, pp 233–246
5. Simmonds J et al (2011) Analyzing methodologies and tools for specifying variability in software processes. Universidad de Chile, Santiago, Chile
6. Reinhartz-Berger I, Figl K, Haugen Ø (2017) Investigating styles in variability modeling: hierarchical versus constrained styles. Inf Softw Technol 87:81–102
7. Bastos JF et al (2011) Adopting software product lines: a systematic mapping study. In: Evaluation & assessment in software engineering (EASE 2011), IET, pp 11–20
8. Santos IS, Andrade RMCC, Neto PAS (2014) How to describe SPL variabilities in textual use cases: a systematic mapping study. In: 8th SBCARS, pp 64–73
9. Lopez-Herrejon RE, Illescas S, Egyed A (2016) Visualization for software product lines: a systematic mapping study. In: IEEE VISSOFT, pp 26–35
10. Sinnema M, Deelstra S (2007) Classifying variability modeling techniques. Inf Softw Technol 49(7):717–739
11. Berger T et al (2013) A survey of variability modeling in industrial practice. In: Seventh international workshop on variability modelling of software-intensive Systems, p 7
12. Petersen K, Vakkalanka S, Kuzniarz L (2015) Guidelines for conducting systematic mapping studies in software engineering: an update. Inf Softw Technol 64:1–18
13. Budgen D, Brereton P (2006) Performing systematic literature reviews in software engineering. In: Proceedings of the 28th international conference on Software engineering, Shanghai, China, pp 1051–1052
14. Kitchenham B et al (2010) Systematic literature reviews in software engineering: a tertiary study. Inf Softw Technol 52(8):792–805
15. Kitchenham BA, Charters S (2007) Guidelines for performing systematic literature reviews in software engineering technical report EBSE
16. Wohlin C (2014) Guidelines for snowballing in systematic literature studies and a replication in software engineering. In: Proceedings of the 18th international conference on evaluation and assessment in software engineering, London, England, United Kingdom, pp 1–10
17. Machado IDC et al (2014) On strategies for testing software product lines: a systematic literature review. Inf Softw Technol 56(10):1183–1199
18. Dyba T, Dingsøyr T (2008) Empirical studies of agile software development: a systematic review". Inf Softw Technol 50(9):833–859
19. Kacper B et al (2016) Clafer: unifying class and feature modeling. Softw Syst Model 15(3):811–845

20. Lanceloti LA et al (2013) Smartyparser: a XMI parser for UML-based software product line variability models. In: Proceedings of the seventh international workshop on variability modelling of software-intensive systems. p 10
21. Passos L et al (2013) Coevolution of variability models and related artifacts: a case study from the Linux kernel. In: Proceedings of the 17th international software product line conference, Tokyo, Japan, pp 91–100
22. Schr R et al (2016) Feature-model interfaces: the highway to compositional analyses of highly-configurable systems. In: Proceedings of the 38th international conference on software engineering, Austin, Texas, pp 667–678
23. Nieke M, Seidl C, Schuster S (2016) Guaranteeing configuration validity in evolving software product lines. In Proceedings of the tenth international workshop on variability modelling of software-intensive systems, Salvador, Brazil, pp 73–80

The Use of Big Data Analytics to Improve the Supply Chain Performance in Logistics Industry

Lai Yin Xiang, Ha Jin Hwang, Haeng Kon Kim, Monowar Mahmood, and Norazryana Mat Dawi

Abstract In recent years, big data analytics has received more attention from companies around the world. The explosive impact of big data analytics on globalized companies has brought new opportunities implementing big data-driven decisions that are sweeping many industries and business functions. Big data analytics has a lot of potential to improve supply chain performance in logistics industry. Big data analytics is frequently used in the logistics/supply chain management industry as an instrument to improve the performance of the system. As the supply chain performance depends on information on a high degree, big data analytics seems to be very useful in improving supply chain performance. However, many companies have not been able to apply to the same degree of the "big data analytics" techniques that could transform the way they manage their supply chains. This research demonstrated how companies can take control of the big data opportunity with a systematic approach. This study utilized survey questionnaires using Google Form to collect data from university students and people working in Klang Valley, Malaysia. It may be viewed that this result unlikely can be an appropriate representation of the whole population of Malaysia. It was concluded that several factors such as improved forecasting, supply chain system integration, human capital and risk and security governance have significant relationship towards supply chain performance in logistics industry.

L. Y. Xiang · H. J. Hwang (✉) · N. M. Dawi
Sunway University, Subang Jaya, Malaysia
e-mail: hjhwang@sunway.edu.my

L. Y. Xiang
e-mail: yinxianggg20@gmail.com

N. M. Dawi
e-mail: norazryana@outlook.com

H. K. Kim
Daegu Catholic University, Gyeongsan, South Korea
e-mail: haengkon@cu.ac.kr

M. Mahmood
KIMEP University, Almaty, Kazakhstan
e-mail: monowar@kimep.kz

© The Author(s), under exclusive license to Springer Nature Switzerland AG 2021
H. Kim and R. Lee (eds.), *Software Engineering in IoT, Big Data, Cloud and Mobile Computing*, Studies in Computational Intelligence 930,
https://doi.org/10.1007/978-3-030-64773-5_2

17

However, two other factors operational efficiency and partner transparency do not have significant relationship with supply chain performance. This research offers a bigger picture of how the use of big data analytics can improve the supply chain performance in logistics industry. Logistics industry could benefit from the results of this research by understanding the key success factors of big data analytics to improve supply chain performance in logistics industry.

Keywords Big data analytics · Supply chain performance · Big data driven decisions

1 Introduction

Supply chain is defined as an integrated process where the raw materials are produced into final products and then delivered to customers either through retail, distribution or both [1, 4]. Distribution, supply management and purchasing has a notable and visible impact on company assets. Many industries are keen to manage the supply chains better. Various methods and technologies such as total quality management, lean production, Just-In-Time (JIT) and agile planning have been implemented to improve supply chain performance. At a strategic level point of view, supply chain management is rapidly emerging and expanding very quickly and it changes the way companies try to meet the demands of their customers.

The explosive impact of e-commerce on traditional brick and mortar retailers is just one notable example of the data-driven revolution that is sweeping many industries and business functions today. Few companies, however, have been able to apply to the same degree the "big analytics" techniques that could transform the way they define and manage their supply chains.

In our view, the full impact of big data analytics in the supply chain is restrained by two major challenges. First, there is a lack of capabilities. Supply chain managers— even those with a high degree of technical skill—have little or no experience with the data analysis techniques used by data scientists. As a result, they often lack the vision to see what might be possible with big data analytics. Second, and perhaps more significantly, most companies lack a structured process to explore, evaluate and capture big data opportunities in their supply chains.

This study is designed to examine how companies can take control of the big data opportunity with a systematic approach. Here, this study analyzes the nature of that opportunity and at how companies in logistics industry can manage to embed data driven methodologies into their supply chain management.

Big data driven supply chain management uses big data and quantitative methods to improve decision making for all activities across the supply chain. In particular, it does two new things. First, it expands the dataset for analysis beyond the traditional internal data held on Enterprise Resource Planning (ERP) and supply chain management (SCM) systems. Second, it applies powerful statistical methods to both new and existing data sources. This creates new insights that help to improve supply chain decision-making, all the way from the improvement of front-line operations, to strategic choices, such as the selection of the right supply chain operating models.

With big data analytics, it is able to determine the behavioral patterns by looking into the data, which also helps in forecasting future behavioral patterns [14]. It is a main driver for capturing business value [4] and has a high potential in setting off a "management revolution" [9].

In spite of all the advantages stated above, Big Data Analytics is still new and deals with a variety of techniques under information science which still in the early years in terms of application in management [2, 9]. As the supply chain performance depend on information on a high degree, big data analytics seems to be very useful in supply chain management context [13]. Data is considered as the element of improved profitability and better decision making. Companies ranked top three in their industry that adopted data driven decision making are averaged 6% more profitable and 5% more productive than their competitors [18]. Therefore, this research has a purpose of having a better understanding on how the application of Big Data Analytics can improve the supply chain performance in the logistic industry. There are many supply chain risks engaging with various companies today, so it is essential to find out and understand how Big Data Analytics could eliminate those risks and provide more lean/agile and efficient supply chains.

The primary purpose of this research is to investigate the use of big data analytics to improve the supply chain performance in logistics industry. To do so, this research was conducted to investigate the relationship between the independent variables such as improved forecasting, operational efficiency and partner transparency and the dependent variable of supply chain performance. Companies will be able to get benefits from this study by understanding and identifying the factors which will improve the overall supply chain performance. Logistics industry managers will be able to reduce unnecessary waste and have a better supply chain performance by adopting Big Data Analytics. Research questions are developed to measure the impact of big data analytics based on key factors such as supply chain integration, improved forecasting capability, operational efficiency, human capital, risk and security governance, and partner transparency.

2 Literature Review

2.1 Big Data Analytics

According to Davenport [2], systematically collecting, analyzing and acting on big data will bring advantage on a corporate level and also a main element to optimize the supply chain structure. This already indirectly saying Big Data Analytics without even pointing out the term. Big Data is special because of the volume, velocity, variety and veracity of data which is commonly available and not expensive to access and store it [9]. Volume can happen in various ways, there are more data because the data are captured in detail. For example, rather than just recording an item sold at a specific location, it also records down the time it sold and the total amount of inventory left will also be recorded. Many companies had started to record daily sales by stock keeping unit and location to make inventory decisions [18].

According to Tan et al. [15], the Big Data is essential for supply chain innovation, the proposed analytic infrastructure method allows firms to consolidate their supply chain innovation and supply chain partners through leveraging the Big Data Analytics insights systematically. An extension in innovation capabilities in supply chain is a great strategy in a fast paced and competitive business environment, considering that important knowledge might reside out of a single firm. This is a new challenge for companies in managing the business and supply chain better by identifying and extracting the most relevant information needed. To analyse Big Data, they need people who are capable and have knowledge to do that job, those people allow businesses, researchers and analysts to execute better decisions by adopting data that they cannot access or cannot use previously. IDC claims that the big data and business analytics worldwide revenue will exceed $203 billion in the year 2020 from $130.1 billion in the year 2016, at a CAGR of 11.7% [7]. This shows that there is a huge potential of applying Big Data Analytics in logistics industry, companies adopting Big Data Analytics are going to be more likely in outrun the competitors in a fast pace and competitive environment.

2.2 Supply Chain Performance

In the current economy, the battlefield is starting to change from individual company performance to supply chain performance. Supply chains are responsible in the whole lifetime of the product, from the preparation of raw materials and supply chain management, to manufacturing and production, distribution and customer service, then ultimately recycling and disposal at the end of the product's life.

Many companies had realized that effective supply chain management played an important role in their day to day operations, and to win in a whole new environment, performance measure is needed for supply chain to have a continuous improvement but most of them failed in building an effective performance measures and metrics required to attain integrated Supply Chain Management. Normally supply chain management experts only review the reduction in cost when they measure the supply chain. But besides cost reduction, there are 3 important metrics to measure the supply chain performance which are inventory measurement, working capital and time [3]. Inventory measurement is important to meet the customers' need while keeping inventory levels at minimum. Lastly, time is important, there are many "time" needed to be measured for example promise time, lead time, cycle time and others. The supply chain performance is greatly dependent on information, the application of Big Data Analytics will bring a lot benefit to it.

2.3 Improved Forecasting

Richards and King [11] argue that the predictions can be push up by using data driven decision making and Tucker [17] is confident that soon big data will be able to predict every move of a consumer more efficiently. According to Richey et al. [12], most respondents see forecasting as the main opportunity for the future, Big data gives a method to assess an expanded set of projections and make strongly informed situations. So, if we got more data, the more analysis can fit in, and that will aid in making right decisions, and then the better we able to execute our strategies". By using Big Data, companies can also provide consumers with what they want in order to satisfy their needs and wants more effective and efficiently than before.

H1   Improved Forecasting is positively related to supply chain performance in logistics industry.

2.4 Supply Chain System Integration

Supply chain system integration can be explained as a close alignment and coordination within a supply chain or in other words, an attempt to bring up the linkages within each element of the supply chain, assist in better decision making and to get all the pieces of the supply chain to interact in more efficiently which directly create supply chain visibility and identify bottlenecks [4, 8]. Nowadays, companies are expected to exploit on Big Data Business Analytics in logistics as well as supply chain operations to boost flexibility, visibility, and global supply chain integration and logistics processes, supervise demand volatility effectively, and handle fluctuation of cost [3, 4]. Furthermore, supply chain analytics enable modelling and simulating complex systems as it focuses on the interrelationship among operations of supply chain and bring attention to the analysis on integral data related to supply chain integration [8]. Big data can improve visibility through linking every supply chain members with the market needs. Besides, adopting integrated information offers more improvement along the supply chain effectively, every party along the supply chain will be benefit from the benefits created by applying big data analytics [12].

H2   Supply Chain System Integration is positively related to supply chain performance in logistics industry.

2.5 Operational Efficiency

Operational efficiency can be explained as a market condition where the participants can carry out transactions and receive services at a price that are considered fair to the actual costs needed to provide them [7]. Big Data Analytics gives advantage as it materializes in continuous optimization, monitoring possibilities and also auto-mated control by analytics-driven real time insights of the supply chain [5]. Applying complex optimization algorithms to the Big Data is a main enabler in making prod-ucts and processes better in terms of consistency and allows the operation of leaner supply chain. According to Richey et al. [12], respondents view the big data as very important in making efficiency in the business and it is useful. If a company has data, they are able to see how their procurement trends are, with it the company will be able to have a strength of bargaining of negotiation, better quality, better lead times, delivery times and efficiency in the plant. These systems will effectively reduce stock out costs and associated with the opportunity costs, it also enables supply chain part-ners to keep track of another partners' capacity and stock. This will help logistics industry to have a leaner supply chain and eliminate unnecessary supply risks.

H3 Operational Efficiency is positively related to supply chain performance in logistics industry.

2.6 Human Capital

According to Davenport [2], data scientists are essential and needed by companies as they are those who are able to understand statistical modelling, operate Big Data systems as well as interpret the data streams. According to Richey et al. [12], data collection process can be improved by recruiting personnel that are knowledgeable about big data. Data scientists enable companies to forecast demand quickly and accurately by empowering the supply chain with predictive analysis. This is important and needs to be made before competitors made the same predictions as well as before customers find out that their desires are not met by your company [6]. Therefore, to improve the supply chain performance in logistics industry, human capital is very important as they are those who analyze the data and provide the results for the company, if the quality of data is good and accuracy is high, the supply chain performance in logistics industry will be improved.

H4 Human Capital is positively related to supply chain performance in logistics industry.

2.7 Risk and Security Governance

According to Richey et al. [12], security governance is important since big data contains a very large amount of information, which is a reason why security is very important. Most of the managers stated that big data security is the main concern particularly as it is related to data storage/accessibility, data ownership and data privileges. Some information is not disclosed with the purpose to protect the organ-isation's solvency. Majority of managers have strong faith that big data analytics can reduce risk related to decision making. After mining big data, top executives are expected to make knowledgeable decisions as well as stakeholders see the information gathered as internally empowering to the whole company as it can reduce risks related with ill-informed decision making.

H5   Risk and Security Governance is positively related to supply chain performance in logistics industry.

2.8   Partner Transparency

Partner transparency can also be explained in the terms of "trust". Increased in transparency allows firms to communicate with each other better, there are many global businesses have at least 3 partners that they are supply chain supplier for, other companies see working with them as an advantage or other particular type of logistics. All they want is to see the transparency of all these different platforms and plans and they can communicate to each other. According to Richey et al. [12], respondents view transparency as a way to get supply chain partners' trust to allow themselves to feel that they can depend on the data shared because the provenance is clear. In a modern supply chain, transparency is essential when doing a business [10]. Availability of end to end real time information access and control is essential for logistic optimization. In order for that to happen, a high level of supply chain transparency and visibility, higher level of visibility lead to a better supply chain efficiency and agility. Every single stakeholder belongs to the value chain will be benefited from transparency not only succeeding value creators but also will go beyond that. This kind of multi-tier visibility enables supply chain decision to be more flexible, dynamic as well as participatory. Partner transparency provides additional data that can help logistic industry in their day to day operations.

H6   Partner transparency is positively related to supply chain performance in logistics industry.

3   Research Methodology

3.1   Research Design

The primary objective of this research is to find out how the use of big data analytics improves supply chain performance in logistics industry. Quantitative method is adopted in this study as it grants a more precise prediction along with the understanding of the topic. Simple random sampling approach is chosen for the research in order to make sure that there exists a high degree of representativeness. It is the efficient form of sampling where every single element in the reachable population has a fair and equal chance of getting chosen. The 7-point Likert scale was used to measure online survey using Google Forms.

Reliability analysis was conducted to make sure that the scale consistently showed the constructs that it is measuring. If the value of Cronbach's alpha rises near to 1, it shows that the items in questionnaire have strong reliability [16].

IBM SPSS Statistics 21.0 Software was used to analyse the data and find out certain relationship between the dependent variables and independent variable To make sure that the data collected from respondents are dependable and accurate, reliability test and standard validity test were conducted. In addition, factor analysis was conducted to confirm the meaningful variables to carry out the research. Multiple regression and ANOVA test were then conducted to perform hypothesis tests.

3.2   Data Collection

The purpose of this research is to identify how the use of big data analytics can improve the supply chain performance in logistics industry. The respondent that were chosen for this research are students and people working in Klang Valley area in Malaysia. Simple Random Sampling was adopted to enable generalizations from the results of a specific sample (252 respondents) back to the whole population (Millennials in Malaysia) due to its high degree of representativeness.

As shown in Table 1, The total of 252 respondents consist of 124 female and 128 male respondents. All 252 respondents are Malaysian and most of them are aged between 18 and 34 years old. The top ages of respondents are between 18 and 24 years old (87.7%) followed by respondents aged 25–34 years old (11.1%). As shown above, majority of the respondents have undergraduate degree as their highest education level followed by postgraduate degree as university is the place where most of them are exposed to big data and supply chain knowledges.

It was also shown in Table 1 that 77.4% of the respondents are students while 13.5% of respondents are full time employed. Table 1 showed that 86.5% of all respondents have a monthly income/ monthly allowance of RM 2500 and below, this is because most of the respondents are universities students.

3.3   Factor Analysis

The Kaiser-Meyer-Olkin Measure of Sampling Adequacy (KMO) and Bartlett's Test of Sphericity were adopted in this research as a form of assumption test where the KMO value is required to be greater than 0.5 in order to determine the adequacy of data. According to the table shown in Table 2, the value of KMO is 0.761 followed by a p-value of 0.000, this shows that the factor analysis conducted is appropriate to explain these data. This suggests that there is a total of 5 separate components extracted by factor analysis.

Table 1 Description of sampled data

		Frequency	Percent	Valid percent	Cumulative percent
Gender					
Valid	Female	124	49.2	49.2	49.2
	Male	128	50.8	50.8	100.0
	Total	252	100.0	100.0	
Age (years)					
Valid	18–24 years old	221	87.7	87.7	87.7
	25–34 years old	28	11.1	11.1	98.8
	Above 34 years old	2	0.8	0.8	99.6
	Under 18 years old	1	0.4	0.4	100.0
	Total	252	100.0	100.0	
Highest education level					
Valid	Postgraduate	34	13.5	13.5	13.5
	STPM/ Pre-U/diploma	4	1.6	1.6	15.1
	Undergraduate	214	84.9	84.9	100.0
	Total	252	100.0	100.0	
Ethnic					
Valid	Chindian	1	0.4	0.4	0.4
	Chinese	209	82.9	82.9	83.3
	Indian	22	8.7	8.7	92.1
	Malay	20	7.9	7.9	100.0
	Total	252	100.0	100.0	
Occupation					
Valid	Employed full time	34	13.5	13.5	13.5
	Employed part time	23	9.1	9.1	22.6
	Student	195	77.4	77.4	100.0
	Total	252	100.0	100.0	
Monthly income/monthly allowance					
Valid	RM 2500 and below	218	86.5	86.5	86.5
	RM 2501—RM 5000	3	1.2	1.2	87.7
	RM 5001—RM 7500	31	12.3	12.3	100.0
	Total	252	100.0	100.0	

3.4 Reliability Analysis

To measure the overall reliability of data gathered, the reliability test was conducted using Cronbach's alpha as shown in Table 3.

Table 2 KMO and Bartlett's test

KMO and Bartlett's test		
Kaiser-Meyer-Olkin measure of sampling adequacy		0.761
Bartlett's test of sphericity	Approx. Chi-Square	5713.551
	df	465
	Sig.	0.000

Table 3 Results of reliability analysis

	Cronbach alpha	No. of items
Improved forecasting	0.716	5
Supply chain system integration	0.778	5
Operational efficiency	0.736	5
Human capital	0.713	5
Risk and security governance	0.788	6
Partner transparency	0.732	5
Supply chain performance	0.807	7

The Cronbach's alpha value for each of the variable is shown in the table above. All variables have Cronbach's alpha value of more than the minimum value of 0.7, therefore those variables were retained for further analysis. Those variables are 5 items for Improved Forecasting (0.716), 5 items for Supply Chain System Integration (0.778), 5 items for Operational Efficiency (0.736), 5 items for Human Capital (0.713), 6 items for Risk and Security Governance (0.788), 5 items for Partner Transparency (0.732) and 7 items for Supply Chain Performance (0.807).

3.5 Pearson's Correlation Coefficient

The correlation between independent variables would render either one of the variables to form a composite variable or to be omitted due to multicollinearity problem if it exceeds a value of 0.7. It was revealed that the correlation between Operational Efficiency (OE) and Improved Forecasting (IF) has a correlation r value of 0.728, Risk and Security Governance (RSG) and Improved Forecasting (IF) has a correlation r value of 0.745, Partner Transparency (PT) and Improved Forecasting (IF) has a correlation value of 0.720, Partner Transparency (PT) and Operational Efficiency (OE) has a correlation value of 0.801, Partner Transparency (PT) and Risk and Security Governance (RSG) has a correlation of 0.704, these values that exceed 0.7 may cause a multicollinearity problem where one of the factors may be redundant.

Multicollinearity test is conducted for the purpose to test the multicollinearity of the constructs, which depicts the tolerance and VIF levels of each variables. If the value of Tolerance levels is above 0.1 and the values of VIF levels is below 10, it shows that there is no significant multicollinearity problem. It was also identified that the tolerance levels for all the above variables are above 0.1 and the VIF levels are less than 10, since there are no violation of the multicollinearity assumption, therefore all the variables used for this research are appropriate.

Model summary was used to find out whether the regression model fits the data or not. According to statistics provided by SPSS, the R Square value is 0.707, which showed that the six independent variables, Improved Forecasting, Supply Chain System Integration, Operation Efficiency, Human Capital, Risk and security Governance and Partner Transparency were able to explain 70.7% of the variation in the use of big data analytics to improve supply chain performance.

4 Findings of the Study

ANOVA was used to test the overall significance of the research model and it showed that F-ratio is 98.346 followed by the p-value less than 0.5, this shows that there is a significant relationship between Supply Chain Performance and the independent variables for this research (Table 4).

As shown in Table 5, the results of hypothesis testing were demonstrated. Four hypotheses were turned out to be failed to reject null hypothesis while two hypothesis were rejected.

H1 Improved Forecasting is positively related to Supply Chain Performance in logistics industry.

The hypothesis proposed for the relationship between Improved Forecasting and Supply Chain Performance found to be significant as it has a p-value of 0.014 where the p-value fulfils the criteria being significant at level where $p < 0.05$. Therefore, H1 is supported.

It is suggested that improved forecasting will have an impact on supply chain performance as there exist a common belief that improved forecasting can improve

Table 4 ANOVA table

ANOVA						
Model		Sum of squares	df	Mean square	F	Sig.
1	Regression	58.988	6	9.831	98.346	0.000[b]
	Residual	24.492	245	0.100		
	Total	83.480	251			

[a]Dependent Variable: SCP
[b]Predictors: (Constant), PT, SCSI, HC, RSG, IF, OE

Table 5 Hypothesis testing

Hypothesis	Relationship	t-statistics	Sig.	Result
H1	Improved Forecasting → Supply Chain Performance	3.613	0.014	Supported
H2	Supply Chain System Integration → Supply Chain Performance	9.088	0.000	Supported
H3	Operational Efficiency → Supply Chain Performance	1.318	0.189	Not supported
H4	Human Capital → Supply Chain Performance	4.454	0.000	Supported
H5	Risk and Security Governance → Supply Chain Performance	3.945	0.000	Supported
H6	Partner Transparency → Supply Chain Performance	0.395	0.693	Not supported

supply chain performance and companies need forecasting abilities to determine efficient inventory levels and customer demands.

H2 Supply Chain System Integration is positively related to Supply Chain Performance in logistics industry.

The hypothesis proposed for the relationship between Supply Chain System Integration and Supply Chain Performance is found to be significant as it has a p-value of 0.000 where the p-value fulfilled the criteria being significant at level where $p < 0.05$. Therefore, H2-is accepted. Supply chain system integration can be explained as an attempt to bring up the linkages within each element of the chain aid in better decision making and to get all the pieces of the supply chain to interact more efficiently which create supply chain visibility and identify bottlenecks [4].

H3 Operational Efficiency is positively related to Supply Chain Performance in logistics industry.

The hypothesis proposed for the relationship between Operational Efficiency and Supply Chain Performance found to not be significant as it has a p-value of 0.189 where the p-value did not fulfil the criteria being significant at level where $p < 0.05$. Therefore, H3 is not accepted. According to Richey et al. [12], respondents view the big data as very important in making efficiency in the business and it is useful. It has the ability to use the information collected to push up all kinds of supply chain partners' "accuracy, efficiency and collaborations". But if there are error in the data, operational efficiency could not be achieved and it might bring more costs to a company. Therefore, this might be a reason why H3 is not supported in this study.

H4 Human Capital is positively related to Supply Chain Performance in logistics industry.

The hypothesis proposed for the relationship between Human Capital and Supply Chain Performance is found to be significant as it has a p-value of 0.000 where the p-value fulfilled the criteria being significant at level where $p < 0.05$. Therefore, H4- is supported. Data collection process can be improved by recruiting personnel that are knowledgeable about big data. Respondents from previous researches stated that there should be a person in—charge of monitoring the data quality, so the person would know if something Is not right and also having a better quality of big data [12].

H5 Risk and Security Governance is positively related to Supply Chain Performance in logistics industry.

The hypothesis proposed for the relationship between Risk and Security Governance and Supply Chain Performance is found to be significant as it has a p-value of 0.000 where the p-value fulfilled the criteria being significant at level where $p < 0.05$. Therefore, H5 is supported.

Most of the managers stated that big data security is the main concern particularly as it is related to data storage/accessibility, data ownership and data privileges. Some information is not disclosed with the purpose to protect the organisation's solvency. Majority of managers have strong faith that big data analytics can reduce risk related to decision making.

H6 Partner Transparency is positively related to Supply Chain Performance in logistics industry.

The hypothesis proposed for the relationship between Partner Transparency and Supply Chain Performance found not to be significant as it has a p-value of 0.693 where the p-value did not fulfil the criteria being significant at level where $p < 0.05$. Therefore, H6 is not supported.

According to Richey et al. [12], respondents view transparency as a way to get supply chain partners' trust to allow themselves to feel that they can depend on the data shared because the provenance is clear. But it is risky as the partner might not be trustable and it will bring a lot of costs to the organisation. Therefore, this may be one of the reasons that H6 is not supported in this research study.

5 Conclusion

Based on the discussions earlier on the use of big data analytics to improve supply chain performance in logistics industry, it could be concluded that some factors such as improved forecasting, supply chain system integration, human capital and risk and security governance have significant relationship towards supply chain performance in logistics industry. However, two other factors, operational efficiency and partner transparency do not have significant relationship with supply chain performance.

Forecasting is generally viewed as the main opportunity for the future as big data provides a method to assess an expanded set of projections and make strongly informed situations. Improved forecasting has the ability to grow the company's ability to address risk related issues and help to make more effective decisions. Supply Chain System Integration helps in better decision making and get all the pieces of the chain to interact in more efficiently which directly create supply chain visibility and identify bottlenecks [4]. Human Capital is also found to have a positive relationship with supply chain performance. To analyse big data, a company needs to have people that have expertise in big data analytics to bring out the full potential of big data. So, the human capital that has expertise in big data analytics can improve the supply chain performance. Risk and Security Governance has a positive significant relationship with supply chain performance. Better risk and security governance will directly improve the supply chain system in logistics industry.

The purpose of this research study was to have a comprehensive understanding of how the use of big data analytics could improve the supply chain performance in logistics industry. Recommendations provided here might be a reflection to solve those limitations and how to improve performance of supply chains in logistics industry. Future researches would be desirable to apply the mixed research method which combines both qualitative and quantitative research engaging with more broad range of respondents to get more relevant and accurate results. This research had offered a bigger picture in how the use of big data analytics can help to improve the supply chain performance in logistics industry. Limitations are observed as most of the respondents are undergraduate students from universities around Klang Valley, Malaysia which unlikely this result can represent the whole target population of Malaysia.

References

1. Beamon BM (1999) Measuring supply chain performance. Int J Oper Manag 2:275–292. Retrieve 10 June 2017 from http://www.emeraldinsight.com/doi/full/10.1108/014435799102 49714
2. Davenport TH (2006) Competing on analytics. Harv Bus Rev 84(1):98–107
3. Genpact (2014) Supply chain analytics. Retrieved 9 October 2017 from http://www.genpact. com/docs/resource-/supply-chain-analytics
4. Hwang HJ, Seruga J (2011) An intelligent supply chain management system to enhance collaboration in textile industry. Int J u- and e-Service, Sci Technol 4(4):47–62
5. IBM (2017) Big Data—IBM analytics. IBM. Retrieved 12 June 2017, from https://www.ibm. com/analytics/us/en/technology/big-data/
6. Intrieri C (2016) Measuring supply chain performance: 3 Core & 10 Soft Metrics. Transportation Management Company | Cerasis. Retrieved 13 June 2017, from http://cerasis.com/2016/ 04/25/measuring-supply-chain-performance/
7. Investopedia (2017) Operational efficiency. Investopedia. Retrieved 12 June 2017, from http:// www.investopedia.com/terms/o/operationalefficiency.asp
8. Kache F, Seuring S (2015) Challenges and opportunities of digital information at the intersection of Big Data analytics and supply chain management. Int J Oper Prod Manag 37:10–36. Retrieved 10 June 2017 from http://www.emeraldinsight.com/doi/full/10.1108/IJOPM-02-2015-0078
9. McAfee A, Brynjolfsson E (2012) Big Data: the management revolution. Harvard Bus Rev 90(10):60–66

10. New S (2015) McDonald's and the challenges of a modern supply chain. Harv Bus Rev February 4, Retrieved 12 June 2017 from https://hbr.org/2015/02/mcdonalds-and-the-challenges-of-a-modern-supply-chain
11. Richards NM, King JH (2013) Three paradoxes of Big Data. Stanf Law Rev Online 66(41):41–46
12. Richey RG, Morgan TR, Hall KL, Adams FG (2016) A global exploration of Big Data in the supply chain. Int J Phys Distrib Logist Manag 49:710–739. Retrieved 9 June 2017 from http://www.emeraldinsight.com/doi/full/10.1108/IJPDLM-05-2016-0134
13. Sahay BS, Ranjan J (2008) Real time business intelligence in supply chain analytics. Inf Manag Comput Secur 16(1):28–48
14. Shmueli G, Koppius O (2011) Predictive analytics and informations systems research. MIS Q 35(3):553–572
15. Tan KH, Zhan Y, Ji G, Ye F, Chang C (2015) Harvesting Big Data to enhance supply chain innovation capabilities: an analytic infrastructure based on deduction graph. Int J Prod Econ 165:223–233
16. Tavakol M, Dennick R (2011) Making sense of cronbach's alpha. Int J Med Educ 2:53–55. Retrieved 17 June 2017, from https://www.ijme.net/archive/2/cronbachs-alpha.pdf
17. Tucker P (2013) The future is not a destination. Available via: http://www.slate.com/articles/technology/future_tense/2013/10/futurist_magazine_s_	predictions_on_quantum_computing_big_data_and_more.html
18. Waller MA, Fawcett SE (2013) Data science, predictive analytics, and Big Data: a revolution that will transform supply chain design and management. J Bus Logist 2:77–84. Retrieved 5 June 2017 from https://pdfs.semanticscholar.org/9c1b/9598f82f9ed7d75ef1a9e627496759aa2387.pdf

Development of Smart U-Health Care Systems

Symphorien Karl Yoki Donzia and Haeng-Kon Kim

Abstract In recent years, the healthcare environment has changed from an existing treatment center to a prevention center and a disease management center. Proposing smart healthcare devices that enable active healthcare by creating a knowledgeable healthcare system or family members rather than single healthcare can effectively expand healthcare services. A family intended for younger users whoa delicacy to an elder family. While the services provided can be used to prevent disease through health care and reduce medical costs, automated care can maximize user comfort and increase demand. In this Work, we are going to develop activating health services to provide personalized services to each user to provide differentiated medical care and change their perception of how to provide services. In this study also, the achievement of service-oriented healthcare information systems focused on the healthcare cloud environment. Therefore, in conjunction with the development of leading smart healthcare systems connected to the IT environment and various smart systems, apparel health management systems emerge by providing patient-oriented services. This study proposes a UML diagram that includes system design to help manage U-Healthcare management problems.

Keywords U-health · Management center · Healthcare · Cloud · Software UML

1 Introduction

Improving the quality of healthcare and improving ease of access to medical records while keeping costs reasonable is a challenge for healthcare organizations around the world. This problem is exacerbated by the rapid increase in the world's population, particularly the growth rate of the elderly will rise to around 1.5 billion by 2050.

S. K. Y. Donzia · H.-K. Kim (✉)
Department of Computer Software, Daegu Catholic University, Gyeongsan-si, South Korea
e-mail: hangkon@cu.ac.kr

S. K. Y. Donzia
e-mail: yoki90@cu.ac.kr

© The Author(s), under exclusive license to Springer Nature Switzerland AG 2021   33
H. Kim and R. Lee (eds.), *Software Engineering in IoT, Big Data, Cloud and Mobile Computing*, Studies in Computational Intelligence 930,
https://doi.org/10.1007/978-3-030-64773-5_3

An aging population means an increase in chronic diseases that require frequent visits. health care providers. As well as higher hospitalization requirements. The cost of treatment increases dramatically as the number of patients requiring continued treatment increases. Over the past decades, information and communication technologies (ICTs) have been widely adopted in health care settings to facilitate access and delivery of health services and make them more cost effective. The use of ICT has led to the development of an electronic medical record (EHR) system. The EHR contains a complete medical record for the patient (current medications, vaccinations, lab results, current diagnoses, etc.) and can be easily shared among multiple providers. They have been shown to improve patient-provider interactions. The adoption of ICTs in the health sector is commonly referred to as digital health. Over the years, digital healthcare has shifted primarily from managing electronic patient data and providing patient web portals to increasing the flexibility and convenience of healthcare management, and is commonly known as connected health. Connected Health uses wireless technologies (Bluetooth, Wi-Fi and Long Term Evolution, etc.) in conjunction with smartphones and mobile apps to allow patients to easily connect with providers without frequent visits.

Connected Health has evolved into smart health using traditional mobile devices (e.g. smartphones) with portable medical devices (e.g. blood pressure monitors, blood glucose monitors, smart watches, smart contact lenses, etc.) and Internet of Things (IoT). Devices (for example, implantable or ingestible sensors) allow you to continuously monitor and treat patients even when they are at home. Smart Health is expected to keep hospital costs low and provide timely treatment for various illnesses by installing IoT sensors in health monitoring equipment. The information collected by these chips can be transmitted to distant destinations. The collected data is sent to the local gateway server through the Wi-Fi network so that the end system can retrieve the collected data from the gateway server. Regular server updates allow clinicians to access patient data in real time. These devices work together to generate integrated medical reports that multiple providers can access. These data are not only useful for patients, but can also be combined to study and predict trends in healthcare across cultures and countries. The amount of data that can be generated through the combination of smart health devices and IoT sensors is huge. This data is often referred to as "big data". Applying effective analytical techniques to big data can deliver meaningful information to clinicians, help them make faster, more informed decisions, and take proactive steps for better health care.

Many frameworks have been proposed to implement IoT technology in healthcare [1]. In the literature on IoT technology, only a few key paradigms have been discussed: application areas, security, and efficiency. The proposed framework focuses on the area of security, efficiency and application of IoT technologies in healthcare. This framework provides a comprehensive design and deployment strategy with IoT technology. It covers all practical aspects of implementing the technology (e.g., communication entities, communication technology, physical structures, data storage, data flow, and access mechanisms). A framework for remote monitoring of the elderly has been proposed. They focused on the applications of IoT technology. His model put forward the needs of the elderly. Machine learning techniques have been used in the proposed model [2]. Monitor the elderly [3]. Swiatek and Rucinski [4] presented the importance of telemonitoring for patients in the applied paradigm. The model they proposed emphasized a decentralized system offering smart health services. They also discussed the innovative and business aspects of eHealth services. Yang et al. [5] He merged the traditional concept of medical boxes with smart health services. An effective system for rehabilitating value for money has been proposed in the IoT environment [6]. The semantic information has been used in the proposed model to efficiently identify the medical resources of the intelligent health system [7]. In the global medical service for the chronically ill and the elderly, there is a growing need for medical information services that can minimize the intervention of experts in terms of cost and treatment time and receive adequate measures for them. symptoms of the disease of the elderly. the patients.. In line with the demands of the time, many research institutes, schools and companies are constantly researching and developing portable computing equipment that can be used in everyday life and is constantly evolving. Among portable computing equipment capable of measuring biological signals, the development of smart clothing capable of measuring various biological signals by placing sensors on the clothing [8, 9]. In particular, due to the development of a digital room capable of communicating, a historic turning point has been established in the development of smart clothing [10]. In the past, smart clothing to measure biometric signals was used by connecting a measurement module or biosignal sensor using various wires or cables. However, since the digital thread is used when sewing directly on the clothes, the weight of the clothes is reduced, and even when used for a long time, there is the advantage of minimizing the rejection of the feeling of wearing.

In Fig. 1. Automation and the frequent intersection of institutional and departmental boundaries Medical companies must integrate systems, data, processes and people while protecting people and personal information. From providing applications to manage medical appointments to providing medical information via the web and supporting critical tests, LSI has extensive experience in the healthcare industry across the globe. They offer a complete portfolio of IT solutions tailored to meet the specific needs of players in the healthcare sector to transform and improve internal processes while ensuring maximum return on technological investments [11]. Their Healthcare provides services in line with market trends current. Concentrate on the supply. Currently, most smart healthcare collects patient information AND analyzes and delivers it, As an incentive, medical care is based on age or disease. It is necessary to provide differentiated services according to the target consumers, etc. Make. In your twenties, there is a lack of awareness of the importance of health care, A systematic and interesting view of health that can motivate The Lee system and the importance of health care for 30–40 year olds is needed Knowing, but becomes indifferent to work and other circumstances Causes procrastination behind me. When and where Insufficient time due to real-time management It would be possible to mitigate areas that may be overlooked and C [12, 13]. Systematic health system for people over 50 Provide health management and remote management by providing adequate health service and in You can reach old age who can have a sense of joy [14].

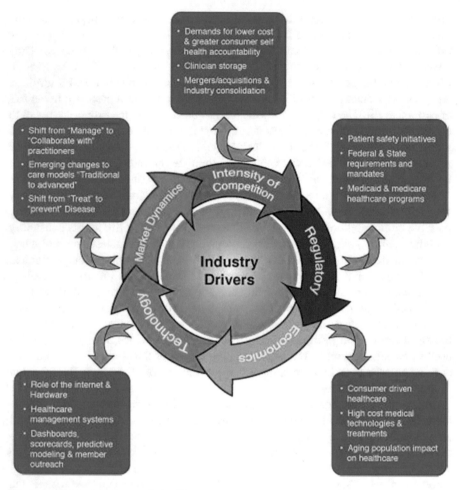

Fig. 1 Technological changes in healthcare

2 Related Literature

Home Reviewing the literature is an important part of any field of study. The reason for this is that it helps researchers identify trends in a particular study and guide research on how to proceed in the area of study. Literature reviews related to studies conducted with GIS were performed. Implementation, accessibility and planning goals are taken into account. In addition to this, it facilitates the identification of research themes and the formulation of research objectives and methodologies. The review of the literature provides a description of the various studies carried out in the field of health, the chosen subject of study. A service. "A review of the literature on the use of GIS-based approaches for medical services": in his article, he attempts to analyze the use of GIS-based measures to explore the relationship between geographic access, use, quality and benefits. health outcomes. The different approaches involved in testing the relative importance of geographic factors that may influence the approach include: inspection. It also focused on a critical assessment of the situation regarding the use of these measures in in-depth accessibility studies using GIS and GPS. However, this review of the literature highlighted the different definitions attached to the term accessibility in the concept of health. Conceptual problems related to the application of GIS. Luis Rosero-Bixby (2004): "Space Access to and Equity in Health Care in Costa Rica: A GIS-Based Study". This article presents a GIS-based analysis. According to the 2000 census, the use of health services by the Costa Rican population. Discuss. To understand health care supply and demand and understand how these two factors converge towards accessibility, the population health services research area sector will monitor and assess the impact of ongoing health reforms. The authors concluded by studying the access card to authorized medical services. Identify geographic disparities and identify specific communities that need help. It demonstrated the need for research areas to optimize location allocation decisions so that health services in general have more equitable organizational access, centrally through the ability to use GIS technology to follow-up. and evaluation. Mark F Guagliard (2004) studied "Spatial access to primary care: concepts, methods and challenges". Analyze basic concepts and measures of access, provide historical context and discuss issues relating to the geographic accessibility of primary primary care, GIS and recent developments in space. Analyze and give examples of promising work. Wei Luo (et al.) (2004): Say "Temporary Changes in Access to Primary Health Care in Illinois (1990–2000) and Policy Implications". In this, the authors studied the temporal change in access to primary care in Illinois between 1990 and 1990. 2000, using the GIS. Census data was used to define population distribution and associated socioeconomic data. It was used to measure non-spatial access. Spatial and nonspatial data were used to access the main areas of scarcity. Through this study, the authors confirmed that spatial access to primary care was the majority of states that improved in 1990–2000. Areas with limited spatial accessibility are mainly concentrated, limited in rural and urban areas, and with socio-cultural disadvantages and barriers, medical needs, mainly due to the population with high socioeconomic scores. Proposes to improve those who need it Population group for the success of future policy. Fahui Wang and Wei Luo (2005) "Assessment of spatial and non-spatial factors of medical access: an integrated approach to health justice In this article, the authors have discussed a spatial approach. Geographic barriers between consumers and providers, and non-spatial factors include non-geographic barriers or enablers such as age, gender, ethnicity, income, social class, education, and skills linguistic. The two-stage floating capture method is implemented in geographic information systems and is used to measure space. Accessibility depending on travel time. Second, the factor analysis method is used to classify several socio-demographic variables into three factors: (1) socio-economic disadvantage, (2) socio-cultural barriers and (3) high medical needs. Finally, spatial and non-spatial factors are integrated to identify areas with poor access to primary care. This study aims to develop an integrated approach to define healthcare professionals. Tribal Areas (HPSA) which can help the US Department of Health and Human Services and the State Department of Health improve the HPSA designation. [15], Johnson and Jasmin Johnson (2001), "GIS-Pandemic Monitoring and Management Tool", In this white paper, the authors attempted to describe the mapping of public health resources, certain diseases and other health events related to their environment.. And with the existing health and social infrastructure, mapping this information creates powerful tools for infectious disease surveillance and management. GIS allows you to use interactive queries for information contained in a map, table or graph. It helps patients to automatically think on the map with dynamic maps published on the Internet. GIS obtains aerial/satellite imagery for easy integration of information such as temperature, soil type and land use, and the spatial correlation between potential hazards and epidemics. Thomas C. Ricketts (2003): "Geographic Information Systems and Public Health". In this article, the author explained that GIS is not complete. It is a solution to understanding the distribution of disease and public health problems, but it is an important way to illuminate how humans interact with the environment or create health. Ellen K. Cromley (2002): wrote an article titled "GIS and Disease". In your review, you use a GIS or remote control to identify and summarize the selected studies. Detected primarily to investigate diseases in the United States. The authors use GIS to describe the distribution of pathogens, how GIS is used to study pathogen exposure, and the important link between environmental and disease surveillance systems and disease-based analysis of disease. GIS. Sara L. Mc. Lafferty (2003): "GIS and HealthCare" study. He analyzed recent literature on GIS and health. Try to analyze in four parts. The first section describes the GIS medical needs analysis study. In the second section, we analyze how GIS is used to study geographic access to health services and to understand the gaps in access between population groups. In the third section, we focus on using [15].

Using GIS to analyze geographic changes in health care use. The last section deals with GIS applications. When evaluating and planning health services. Support systems for spatial determination and modeling of location assignments were also discussed. He concluded that GIS offers a new way to study small medical needs. Geographic area, better measurement and analysis of geographic access to health care and location of planning services. These new spatial behaviors that can be studied and modeled in GIS will give priority to future research interests. Sabasen S and Raju K. H. K (2005), "GIS for health and sustainable development in rural India, special reference to vector-borne diseases", in this article. The authors discuss how to approach advanced knowledge of the basics of disease transmission. Satellite epidemiology, environmental factors and disease outbreaks can be predicted Remote sensing data reviewed for modern tools such as remote sensing geographic information systems (GIS) are suitable for solving disease surveillance, control, monitoring and evaluation problems. Information systems designed in the GIS field of rural medicine, this article will explain how it will ultimately facilitate their use. Ensuring resources, disease prevention and health promotion, general rural development and, therefore, program maintenance at all levels. AnkitaMisra, (et al.), (2005): Wrote a joint article on "GIS analysis of healthcare in Pune". They are trying to use GIS to analyze the distribution of hospitals and illnesses, availability and use of health facilities in Pune region. The authors believe that local health care It was well stocked, but in some areas of the study area the roads lacked hospital services and recommended areas where possible for establishing new hospitals in Pune district. Shanon et al. (1973) The use of health services was analyzed mainly in terms of access, in terms of geographical distance and time. Access to medical services is linked to the distance from the center and the time it takes for patients to get to the center. Ronald Anderson and John. F. Newman (2005) argued that in "The Social and Individual Determinants of Health Care in the United States", health use is underscored by characteristics of the health care delivery system, changes in medical technology and the determinants of health. Disposable. The author has these three factors Health insurance system. Roger Strasser (2003): "Rural health in the world: challenges and solutions". In this article, the author has discussed and reviewed the issues facing the main challenges facing rural health around the world and attempted to analyze that enormous challenges have been met to improve the health of populations. rural and remote areas around the world., which include: A concrete action plan has started., A global initiative for rural health planning. Rural health care around the world to provide "all rural health" to all rural people through the intense efforts of international and national workers such as doctors, nurses and other health workers in rural areas of the whole world. Center. The health and well-being of people living in rural and remote areas of the world usher in a new era of improvement. Hilary Graham and Micheal, P. Kelly (2004) The authors of this article "Inequalities in health: concepts, frameworks and policies" have attempted to highlight the conceptual problems associated with socio-economic inequalities in health. In the first section, people Filed in the UK and how the issue of health inequalities is addressed using traditional measures of socioeconomic status. The second section focuses on "determinants," a key term in the drive to reduce health inequalities, and examines the difference between the determinants of health and the determinants of health. Inequality in health. The third section separates the concepts of health disadvantage, health disparity and health gradient. Therefore, this thesis clarifies some of the key terms used in the discussion of health inequalities to help inform the policy-making process. Tulio Zappelli, Luigi Pista Perry, Guglielmo Weber (2006) worked on "Quality of health, economic inequalities and preventive economies", in this article the author analyzed the overall variability of the quality of medical care provided by the healthcare system. universal public health of Italy to determine its effectiveness.. Questions on income inequality and health inequalities AND preventive savings. They found that there was more dispersion of income and health and more preventive savings in areas of lower quality. This analysis provides important information on the ongoing debate on the feasibility of life cycle models and the interesting policy implications for the design of health systems.

3   Background of the Study

3.1   Objectives of the Study

Since Health is a fundamental area to be developed for a better standard of living and involves the treatment and management of diseases. Preserving health in the use and connection of the medical professions through the services provided by laboratories of medical, dental, pharmaceutical and clinical science (e.g. diagnostics). The main theme of health care is to provide comprehensive health facilities, protect all physical, social and mental health, reduce mortality, increase human life expectancy, and develop socially for balanced development. The first reason is that it helps researchers identify trends in a particular study and guides them in their research on how to proceed in their area of study. We use advanced GIS tools to enable researchers and planners to visualize and conceptualize health plans and policies. In this article, we will show you how to combine Internet of Things (IoT) and Big Data technologies with smart health to deliver better health solutions in detail, using UML software methodology effectively, in case diagrams of use, class diagrams and sequence diagrams. Try to deliver by design.

3.2   Research Survey

The main objective of this white paper is to present a special framework for smart health systems based on IoT. The framework uses a layered approach to address key IoT-based smart health system challenges and provides a comprehensive mechanism for collecting patient data to cloud storage. The model shown also uses limited application protocols and hypertext transfer protocols in a sandbox. Different requirements of the two environments in the Internet environment. At the local public health center, a certified medical specialist Use a list verified by health authorities. This approach In our study, we only guarantee qualified professionals from local health centers. Web technology for the IoT. It can be used to develop existing web power generation IoT systems. However, these advances are not enough to adequately support IoT systems. The result is therefore not good. You can pass JSON and XML as payloads using HTTP and WebSocket protocols. Traditional web protocols can be implemented on IoT devices, but these protocols require more resources to support IoT applications. To support IoT systems, many specialized protocols have been developed that can work effectively on resource-constrained IoT devices and networks [16].

Several research issues have been raised in the deployment of IoT technology. Various work has been done in the research community to alleviate the problem and find a better implementation solution. Other frameworks in the field of health sciences are It has been proposed to efficiently distribute smart health services.

The model presented also uses a limited application protocol in a sandbox and a hypertext transfer protocol in an Internet environment because the requirements of the two environments are different.

3.3   Embedded Software Development Life Cycle

In general, VMware is virtual and cloud computing software. VMware is a subsidiary of Dell Technologies and is based on virtual bare metal technology. The VMware application system in Fig. 2 is not included in the study analysis, but is used to demonstrate the importance of software methodology. Analysis, design, implementation, testing and maintenance are very relevant to our research. SDLC (Software Development Life Cycle) is the process of creating or maintaining a software system [6]. This usually involves a series of steps, from preliminary development analysis to testing and evaluating post-development software. It also consists of models and methodologies used by development teams to develop software systems, which form a framework for planning and controlling the entire development process. The SDLC selection and adoption process is essential to maximize the possibility for organizations to successfully deliver their software, so choosing and adopting the right SDLC is a long-term management decision [17].

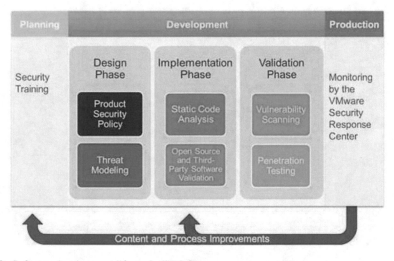

Fig. 2   Software development life cycle (SDLC)

4 Overall Architecture of Smart U-Health Care Systems

UML (Unified Modeling Language) is a standard visual model.

Language used in the analysis, design and implementation of software systems. UML is a way to visualize software using a collection of diagrams. The unified modeling language UML aims to become the industry standard for software development and documentation. But the research project also UML with a lot of effort. By describing UML in a simplified way, we can say that the language provides a set of different diagrams to highlight different aspects of the software. Different diagrams are typically used differently at different stages of software development.

Implementation solution which are detail on Figs. 3 and 4.

4.1 Use Case Diagram

Use cases illustrate the units of functionality provided by the system. The main objective of the use case diagram is to assist in the development of applications for health in the use and connection of health professionals through the services

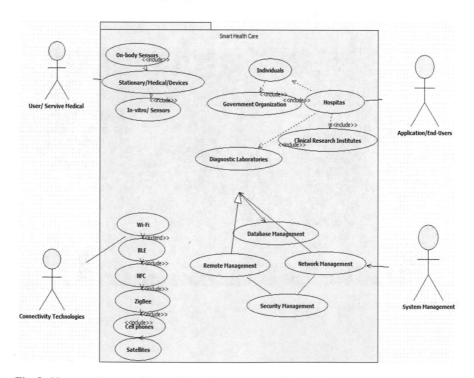

Fig. 3 Use case diagram of Smart U-health care systems (1)

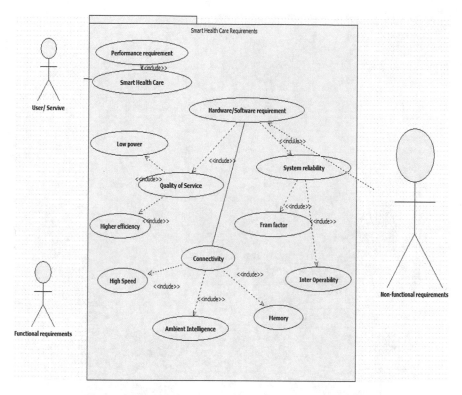

Fig. 4 Use case diagram of Smart U-health care systems (2)

provided by the medical, dental, pharmaceutical science laboratories and clinics. Use cases illustrate the units of functionality provided by the system. Including "user" relationships. Use case diagrams are generally used to convey the high level of functionality of our system. [The theme of health care is to provide comprehensive medical facilities. You can easily perform the functions provided by the sample system [17].

4.2 Class Diagram

The class diagram shows "Architecture of Smart U-health care Systems entities which are a connection of health professionals through the services provided by the medical, dental, pharmaceutical science laboratories and clinics as showed in Figs. 5 and Fig. 6

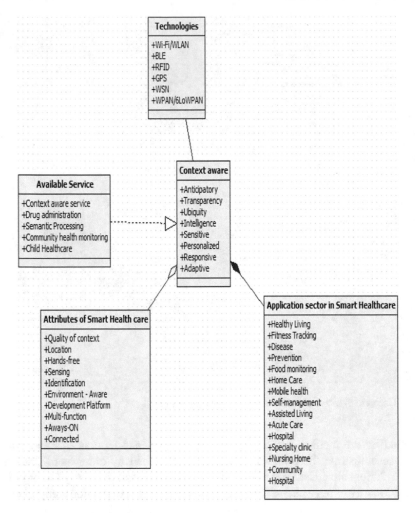

Fig. 5 Class diagram of Smart U-health care systems (1)

4.3 Sequence Diagram

The Sequence diagrams showing Fig. 7 detailed flows for specific use cases that facilitate the functionality of the app once animals are detected. Demonstrate the high level of functionality of our sequence, as can be seen in Fig.7.

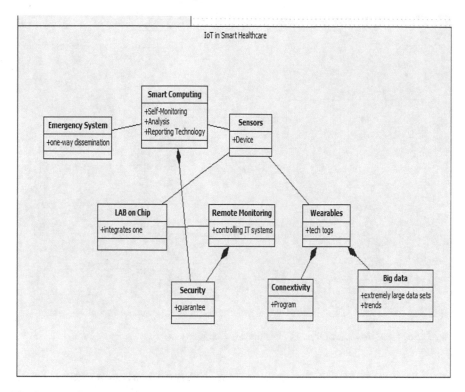

Fig. 6 Class diagram of Smart U-health care systems (2)

5 Conclusion

In this study, in order to provide differentiated medical services and to change the perception of the mode of service delivery, we develop an active health service that can provide personalized services to each user, and in the architecture design by diagram UML software, service-oriented healthcare. There are effective results. Information system focused on the health cloud environment. The literature review was organized in a very systematic order. It helped me understand the definition of healthcare, MIS and healthcare, healthcare inequalities and healthcare utilization. This framework studies various aspects of IoT technology for smart health services, such as the internet environment. Communication protocol and web technical requirements. The proposed model consists of 3 layers, each of which performs a special task. Going forward, our goal is to develop a detailed system architecture design security infrastructure using UML diagrams and real-time system applications that can be integrated using the current framework.

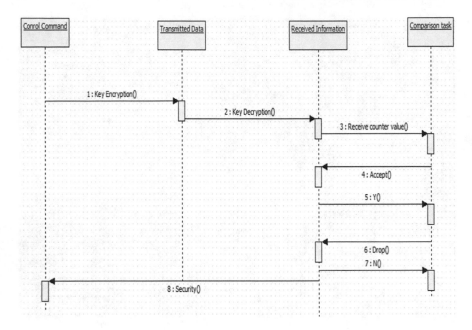

Fig. 7 Sequence diagram of Smart U-health care systems

Acknowledgements "This work was supported by research grants from Daegu Catholic University in 2020".

References

1. Anselin M, Sahi MA, Abbas H, Saleem K et al (2017) A survey on privacy preservation in E-healthcare environment. IEEE Access 2017
2. Earley S (2015) Analytics, machine learning, and the internet of things. IT Prof 17(1):10–13. Article ID 7030173
3. Basanta H, Huang P, Lee TT (2016) Intuitive IoT-based H2U healthcare system for elderly people. In Proceedings of the networking, sensing, and control (ICNSC), 2016 IEEE 13th international conference on, IEEE, pp 1–6, 2016
4. Swiatek P, Rucinski A (2013) IoT as a service system for eHealth. In Proceedings of the E-health networking, applications & services (Healthcom), 2013 IEEE 15th international conference on, IEEE, Lisbon, Portugal, pp. 81–84, October 2013
5. Yang G, Xie L, Mantysalo M et al (2014) A health-IoT platform ¨ based on the integration of intelligent packaging, unobtrusive bio-sensor, and intelligent medicine box. IEEE Trans Industr Inf 10(4):2180–2191
6. Fan YJ, Yin YH, Xu LD, Zeng Y, Wu F (2014) IoT-based smart rehabilitation system. IEEE Trans Industr Inf 10(2):1568–1577
7. Pasha M, Shah SMW (2018) Framework for E-health systems in IoT-based environments. Wirel Commun Mob Comput 2018, 11 p. Article ID 6183732
8. Kim SH, Ryoo DW, Bae C (2008) U-healthcare system using smart headband. In: Proceeding of 30th annual international IEEE EMBS conference, pp 1557–1560, 2008

9. Oh SJ, Lee CW (2008) u-Healthcare sensor grid gateway for connecting wireless sensor network and grid network. In: Proceeding of international conference on advanced communication technology, pp 827–831, 2008
10. Chung GS, An JS, Lee DH, Hwang CS (2006) A study on the digital yarn for the high speed data communication. In: Proceeding of The 2nd international conference on clothing and textiles, pp 207–210, 2006
11. http://www.lsioutsourcing.org/LSI/data/industries/life-sciences.php
12. Raquel F, Adriano C (2017) Development of Android-Mobile Application Software in Teaching Web System and Technologies. International Journal of Emerging Multidisciplinary Research, 1(1):53–61. https://doi.org/10.22662/IJEMR.2017.1.1.053
13. Moon HY (2012) Macrophage migration inhibitory factor mediates the antidepressant actions of voluntary exercise. Proc Natl Acad Sci 109(32):13094–13099. https://doi.org/10.1073/pnas.1205535109
14. Weiss G (2000) A modern approach to distributed artificial intelligence. IEEE Trans Syst Man Cybern-Part Appl Rev 22(2)
15. Minutha Minutha V, A literature survey on health care service. University of Mysore Mysore, Karnataka, India, See discussions, stats, and author profiles for this publication at: https://www.researchgate.net/publication/305473861
16. Pasha M, Shah SMW (2018) Framework for E-health systems in IoT-based environments. Department of Information Technology, Bahauddin Zakariya University, Multan, Pakistan. Wireless Communications and Mobile Computing Volume 2018, Article ID 6183732, 11 p
17. Donzia SKY (2019) Development of Predator Prevention Using Software Drone for Farm Security. Department of Computer Software. Daegu Catholic University Thesis 2019.022

An Implementation of a System for Video Translation Using OCR

Sun-Myung Hwang and Hee-Gyun Yeom

Abstract As the machine learning research has developed, the field of translation and image analysis such as optical character recognition has made great progress. However, video translation that combines these two is slower than previous developments. In this paper, we develop an image translator that combines existing OCR technology and translation technology and verify its effectiveness. Before developing, we presented what functions are needed to implement this system and how to implement them, and then tested their performance. With the application program developed through this paper, users can access translation more conveniently, and also can contribute to ensuring the convenience provided in any environment.

Keywords Machine learning · Optical character recognition · Image translator · Machine translation · Video translation

1 Introduction

After starting with the AI Go, which defeated the world Go, the topic of "artificial intelligence" began to emerge as a topic of interest in the field of interpretation and translation. In the near future there was anxiety that machines might replace humans in most translations, and the translation industry was also watching the 4th industrial revolution. With the advent of new technologies, life, work, and communication methods are changing rapidly and unprecedentedly, and expectations and anxiety in the field of artificial intelligence are crossing.

Remarkable changes in the translation area between different natural languages using machines are optical character recognition technology and translation technology [1–4]. Optical Character Recognition (OCR) technology that improves the

S.-M. Hwang · H.-G. Yeom (✉)
Daejeon University, Daejeon, South Korea
e-mail: yeom@dju.ac.kr

S.-M. Hwang
e-mail: sunhwang@dju.ac.kr

© The Author(s), under exclusive license to Springer Nature Switzerland AG 2021
H. Kim and R. Lee (eds.), *Software Engineering in IoT, Big Data, Cloud and Mobile Computing*, Studies in Computational Intelligence 930,
https://doi.org/10.1007/978-3-030-64773-5_4

49

recognition rate of characters by learning numerous character patterns compared to the phrase-based machine translation (PBMT) method, in which the language to be translated in the past was translated by character units Neural Machine Translation (NMT) technology, which stores and learns a lot of sentence information and translates it, is an advanced technology, and its efficiency has been greatly improved. A representative example of this is the Google Translation application. Through this, the OCR function as well as the translation is supported, so that the user can receive the translation by focusing on the desired screen anywhere. However, unlike the mobile environment where the OCR function was supported, the translation function is not supported in the desktop environment. In particular, this problem cannot be provided with sufficient service in situations where long text images such as PDFs or subtitles requiring numerous repetitive translations are required [5–7].

In order to solve this problem, this paper aims to expand the user's convenience by developing an image translator program that extracts text by decoding a desktop-based image by integrating existing OCR functions and translation technology and translates the extracted text.

2 Translation System

2.1 Google Cloud Vision API

Google Cloud Vision API is an image analysis service provided by Google and operates in the cloud. It provides the ability to recognize individual objects in an image and classify them into thousands of categories using machine learning models, or to monitor various types of inappropriate content, from adult content to violent content. Also, because it provides REST API, users can provide information on image files locally or remotely, and it can receive meta files in JSON structure, making it easy to extract and process information. In addition, the Vision API has the advantage of using OCR to detect text in over 50 languages and various file formats in an image. Table 1 shows service information provided by Google Cloud Version API. The OCR service detects and extracts text (text) from an image. For example, a picture with a sign or a sign. JSON contains individual words and their bounding boxes along with the entire extracted string.

The image attribute detection API is a function that detects general attributes such as the main color of an image. The face detection API is a function that detects multiple faces in an image together with key face-related properties such as an emotional state. Facial recognition is not supported. The logo detection API is a function that detects a popular product logo in an image. The label detection API can identify objects, places, activities, animal species, and products through labels.

Table 1 Google Vision API Feature

Feature	Description
Optical character recognition	Detect and extract text within an image, with support for a broad range of languages, along with support for automatic language identification
Image attributes	Detect general attributes of the image, such as dominant colors and appropriate crop hints
Face detection	Detect multiple faces within an image, along with the associated key facial attributes like emotional state or wearing head wear
Logo detection	Detect popular product logos within an image
Label detection	Detect broad sets of categories within an image, ranging from modes of transportation to animals
Explicit content detection	Detect explicit content like adult content or violent content within an image
Landmark detection	Detect popular natural and man-made structures within an image
Web detection	Search the internet for similar images

2.2 Neural Network-Based Machine Translation Model

The representative natural language processing field to which deep learning technology is applied can be said to be machine translation. Neural network-based machine translation (NMT) is a machine translation based on several existing modules in that the translation model is constructed and trained as a neural network [8–10]. In general, the NMT is composed of an encoder and a decoder, and the encoder expresses an input sentence composed of words in a vector space, and the decoder creates the words of the output sentence one by one in sequence. This process is different from the traditional machine translation system dealing with words at the symbol level [11].

2.3 Tesseract

Tesseract is an open source OCR engine developed by Hewlett and Packard (HP) for about 10 years since 1984. In 2005, Tesseract was released as open source through continuous performance improvement after development. Google now supports part of Tesseract [12–15].

Since Tesseract does not analyze the page layout, the image with the text area set must be input [5]. Tesseract is used as part of the OCR server and performs the main function of extracting text.

2.4 *WeOCR*

WeOCR is a platform of Web-enabled OCR system that enables character recognition on the network [12–15]. WeOCR receives the image file from the user, recognizes the text in the image, extracts it, and delivers it to the user. WeOCR serves as a toolkit for convenient use of the OCR engine on the web and supports interworking with open source OCR engines such as Tesseract, GOCR, and Ocrad.

3 System Structure

3.1 Translation Process

The proposed system is largely composed of the original image and then saved as an image, and then is processed through the image processing process, the recognition process, and the translation process, and is composed of the final output process. will be displayed on the screen. Figure 1 is a flow chart of the proposed system.

It is a video translation system that converts the image to the image by specifying the translation area on the video screen, and then extracts the text from the image and translates it to output. The user designates an image of the translation screen area on the video screen and checks the translation result.

First, the image segmentation process extracts a border group recognized as a character from the translation screen image, and filtering removes a non-text area to

Fig. 1 Flowchart of the proposed system

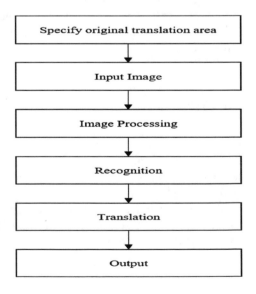

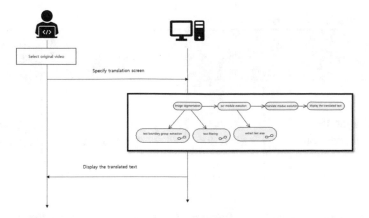

Fig. 2 System architecture

in-crease the text recognition rate of the border group. The OCR module recognizes the extracted text through image segmentation and extracts the correct text area.

The translation module translates into the designated language using the extracted text. The screen output displays the translated text in a designated area. Figure 2 is the structure of the proposed system.

3.2 Main Function

As shown in Fig. 3, there are three main functions of the use case model, which are divided into the text area extraction function, the text translation function, and the text output function that you want to translate from the first original video.

Fig. 3 Use case model

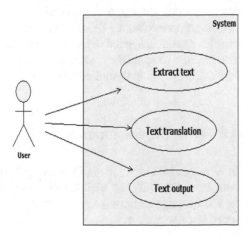

Table 2 Boundary extraction process

Step 1	Step 2	Step 3	Step 4
Grayscale transformation	Outline extraction	Apply threshold	Close operation

text extraction function is a function that extracts a text area through image segmentation by specifying a translation screen area in the original video, and ex-tracts accurate text by performing an OCR function.

text translation function is a function that receives the text extracted from the text extraction function and is translated using the translation API module.

text output function is a function that receives translated text and outputs it to the original designated area. The translated text is processed in a Blur on the existing screen to prevent overlapping with the existing text.

4   Function Implementation and Evaluation

4.1   Translation Process

The image segmentation function was implemented with the OpenCV library. As shown in Table 2, text is extracted through the image segmentation process. The first converts a color image to a grayscale image. The second performs morphological calculation (Morphology Ex) of expansion and erosion to extract the outline. Third, a threshold value is applied to the processed image to obtain a black and white image. The fourth is to apply the erosion operation to the image to which the morphological expansion operation is applied again to emphasize the boundary of the image.

4.2   OCR Implementation

In Windows PC environment, OCR, or optical character recognition, is implemented with Tesseract-OCR and Vision API library. As a step of extracting the text existing on the image, two libraries are used as a method to increase not only processing speed but also accuracy.

The Vision API is a function that analyzes an image, not just a character. It not only has a high recognition rate of characters on the screen, but also provides various in-formation such as location coordinates of characters and language of characters.

4.3 Implementation of Text Translation

Text translation was implemented with the Google Translation API. Google Translate uses the NMT method, and learns and analyzes the entire sentence, provides the context of the entire sentence, and corrects unstable translations.

4.4 Text Output Implementation

To output the translated sentence on the display, Blur processing was performed on the existing screen to prevent overlapping with the existing text. Blur processing on the existing screen prevents overlapping with existing text, so natural output on the existing screen is possible.

4.5 Interface Implementation

The window-based interface is Fig. 4a. Figure 4 shows sequentially from the main execution screen of the window-based proposal system.

Figure 4a shows the menu. If you click the 'Capture' button, cut the text from one image in Fig. 4b and click the 'Language' button. Figure 4c shows the result of text translation output.

4.6 System Evaluation

The expected time of the proposed system is expected to be about 1–2 s, and the text recognition accuracy is expected to be 98%. The main advantage of this system is convenience and versatility. In the Windows PC environment, the user can designate the desired image area and receive the translation. Even if a separate out-put area is not allocated, it can be directly checked on the existing screen. Also, since the video is analyzed and translated as it is, translation is possible in any environment where text is present. Table 3 shows the evaluation with the existing system. The proposed system has a slow output speed but a high character recognition rate. While the existing system output is in a separate output area, the proposed system outputs in

Fig. 4 System user interface

(a) Main Screen

(b) Text extraction   (c) Text output

Table 3 System evaluation

Evaluation item	Existing system	Suggestion system
Character recognition rate	Middle	Height
Translation rate	Middle	Height
Output speed	0.5–1 s	1–2 s
Output method	Other area	Original area

the text area of the original text area. Therefore, more natural translation results can be confirmed.

5   Conclusion

In this paper, we developed an image translation system that combines OCR technology and translation technology.

When the user executes the program, the translation is provided through a series of processes. After setting the desired area, the process goes to the screen recognition step, and after the text position analysis is completed, the process goes to the OCR step. The text identified by the community OCR is translated and output to the designated display area in the text output step. This provides great convenience to the user, and also provides versatility that can be used anywhere.

Acknowledgements This research was supported by the Daejeon University Research Grants (2019).

References

1. Cho K et al (2014) Learning phrase representations using RNN encoder-decoder for statistical machine translation. arXiv preprint arXiv:1406.1078
2. Bahdanau D, Cho K, Bengio Y (2014) Neural machine translation by jointly learning to align and translate. arXiv preprint arXiv:1409.0473
3. Tu Z et al (2017) Context gates for neural machine translation. Trans Assoc Comput Linguist 5:87–99
4. Vaswani A et al (2017) Attention is all you need. Adv Neural Inf Process Syst
5. Ma M et al (2017) Osu multimodal machine translation system report. arXiv preprint arXiv: 1710.02718
6. Madhyastha PS, Wang J, Specia L (2017) Sheffield multimt: using object posterior predictions for multimodal machine translation. In: Proceedings of the second conference on machine translation
7. Caglayan O et al (2017) Liumcvc submissions for wmt17 multimodal translation task. arXiv preprint arXiv:1707.04481
8. Kalchbrenner N, Blunsom P (2013) Recurrent continuous translation models. EMNLP
9. Sutskever I, Vinyals O, Le QV (2014) Sequence to sequence learning with neural net-works. Adv Neural Inf Process Syst (NIPS)
10. Bahdanau D, Cho K, Bengio Y (2015) Neural machine translation by jointly learning to align and translate. In: International conference on learning representations (ICLR), 2015
11. Koehn P (2010) Statistical machine translation. Statistical machine translation. Cambridge University Press. ISBN 9780521874151
12. Mithe R, Indalkar S, Divekar N (2013) Optical character recognition. Int J Recent Technol Eng 2:72–75
13. Go EB, Ha YJ, Choi SR, Lee KH, Park YH (2011) An implementation of an an-droid mobile system for extracting and retrieving texts from images. J Digit Contents Soc 12(1):57–67
14. Cho MH (2010) A study on character recognition using wavelet transformation and moment. J Korea Soc Comput Inf 15(10):49–57
15. Song JW, Jung NR, Kang HS (2015) Container BIC-code region extraction and recognition method using multiple thresholding. J Korea Inst Inf Commun Eng 19(6):1462–1470

Research on Predicting Ignition Factors Through Big Data Analysis of Fire Data

Jun-hee Choi and Hyun-Sug Cho

Abstract The National Fire Agency collects data such as reporting, dispatch, and suppression of fire incidents in the country every year. However, it remains at the basic level of analysis of frequency or cause of fire that occurs. In response, about 46,000 fire data throughout the country were analyzed in 2018 to predict the cause of ignition in the event of a fire. The majority of data recorded in a single fire event is unsuitable for analyzing a fire, and there is no value or data irrelevant to the cause of the fire. Thus, the data was refined to about 30,000 cases, excluding meaningless values. In addition, because it is a study to infer the unknowns of the ignition factors, data that are directly related to the ignition factors were excluded. As a result, an artificial neural network algorithm was applied to infer the ignition factors using about 10 data per fire accident, and the prediction accuracy was about 80%. Rather than determining the ignition factors through this data analysis, it is expected to provide information to fire extinguishing groups and help to increase the level of fire extinguishing. In addition, in order to use the collected data for analysis and prediction, the structure of the database needs to be improved. If the system is improved with artificial intelligence in fire inspection, it is expected that the analysis results of artificial intelligence will be provided in real time by inputting information at the site.

Keywords Fire safety · Fire protection · Disaster prevention · Fire suppression · Fire science

J.H. Choi (✉)
Department of Disaster Prevention, Graduate School, Daejeon University, Daejeon, South Korea
e-mail: junhee3020@hanmail.net

H.-S. Cho
Department of Fire and Disaster Prevention, Daejeon University, Daejeon, South Korea
e-mail: chojo@dju.kr

1 Introduction

According to domestic data, about 46,000 fires occur every year. About 200 types of data such as date and time, dispatch time, weather, and cause of fire are recorded for each fire. Based on these data, the National Fire Agency prepares statistical data such as regional, seasonal, and fire factors, or utilizes them as fire prevention and analysis data. In practice, however, rather than being used as data to prevent fires, it remains at a simple level of statistics and analysis, such as the frequency of fires by location or major causes by season.

When a fire inspection team records the cause after a fire is extinguished, the results of the fire factors may vary depending on the experience of the fire inspection team. Fires have various causes such as mechanical, electrical, and natural factors, but only a few fire inspectors have expert knowledge in all fields. In addition, if the site is burned down due to a delay in the arrival of the fire fighters or a fire is reported a little late, it may be difficult to determine the initial cause, and a fire with an unknown ignition factor may occur for various reasons.

To minimize the influence of fire inspection experience or additional situations, fire ignition factors were inferred using artificial intelligence algorithms and past fire data. Research in the field is in its early stages and is not far enough along to replace human work with artificial intelligence algorithms, but it is expected to help to identify the ignition factors of fire based on data given from the site with a fire extinguished.

2 Related Work

Research to predict fire or reduce damage caused by fire is being conducted from various perspectives. The research is largely divided into two directions, one that allows firefighters to quickly enter the site when a fire occurs, or the other that uses data to analyze and predict fires.

Although there have been cases of research on predicting fires by analyzing big data, it is difficult to predict the cause of fires because they have various causes. Among the analysis methods for specific areas, there is a study that collected and analyzed fire data from Jinju City [1]. This research provided a foundation for establishing a plan to minimize damage and respond quickly when a fire broke out by linking fire information and location information that occurred in Jinju City.

In addition, fire data from Jinju City were used for analysis, providing information on areas or buildings where fires frequently occurred [2]. Regression analysis and spatial correlation were used to list results strongly related to the occurrence of fire [3].

In some cases, there are studies that analyze the prediction and cycle of conflagration fire accidents using classification algorithms or classification using the expertise of firefighters [4]. In the case of the clustering algorithm, there was a limit of high deviation, but the result of a certain pattern was derived.

In recent years, there have been studies on analyzing fire trends and standardizing fire factors using big data [5–11], and data from various perspectives, such as fire data collected by the National Fire Agency every year or decisions made by firefighters, are being utilized.

3   Fire Factor Prediction and Analysis

3.1   Dataset

The National Fire Agency collects data that on fires occur in the country every year. In this paper, the ignition factors were estimated by using data generated throughout the country in 2018. A dataset was made consisting of a total of more than 200 columns, but sometimes it overlapped with the hour column and the second column. In addition, most columns were not suitable for data analysis due to scarcity where data did not exist. Therefore, the data was preprocessed to be suitable for analysis (Table 1).

In order not to face scarcity problems in the process of selecting data to analyze ignition factors, columns with a high percentage of data present in each fire were selected as priorities. Information on fire locations and fire stations that were not related to ignition factors were excluded. Small and medium classifications were considered unsuitable for prediction due to their great diversity of data, and text data in a descriptive form was excluded.

Columns such as ignition heat sources, which are directly related to ignition factors, were excluded because they had a high proportion of unknown values if the ignition factors were unknown. The purpose of the study was to predict unknown causes of ignition. However, because supervised learning must know the value of the ignition factor, the experiment excluded unknown data from the ignition factor.

About 25,000 data were used for the analysis of ignition factors, and the experiment was conducted with about 20,000 training data and 5000 test data. After preprocessing the data, the sparsity of the data used in the experiment was 9.15%, and most of the data had values. Sparsity refers to the ratio of empty values in the total data.

3.2   Performance Index

The performance indicators used the precision of matching the values of the actual ignition factors with the values of the predicted ignition factors. In the experiment, it means the proportion of the matched number of nine resultant values of the ignition factor. The precision has a value from 0 to 1, and is 0 when all predictions fail, and 1

Table 1 Sparsity and duplicate columns

No	Category	Column	Format
1	Sparsity column	Car location	Text (6 types)
2		Car	Text (2 types)
3		Country of manufacture of ship or aircraft	Text
4		Product name	Text
5		Product number (unique number)	Text
6		Manufacturing date (year)	Text
…		Etc.	
1	Date and time	Date of fire occurrence	YYYY-MM-DD HH:MM:SS (AM or PM)
2		Fire occurrence (year)	YYYY
3		Fire occurrence (month)	MM
4		Fire occurrence (day)	DD
5		Fire occurrence (hour)	HH (24)
6		Fire occurrence (min)	MM
7		Day of the week	Text
8	Fire suppression	Call date and time	YYYY-MM-DD HH:MM:SS (AM or PM)
9		Dispatch date and time	
10		Arrival date and time	
11		Initial date and time	
12		Fire suppression date and time	
13		Return date and time	
14		Dispatch time	HH:MM:SS (AM or PM)
15		Fire suppression time	HH:MM:SS (AM or PM)
16		Dispatch time (h)	HH (24)
17		Dispatch time (min)	MM
18		Dispatch time (s)	SS

when all predictions are successful. The artificial intelligence algorithm used in the experiment has nine outputs. Thus, it is possible to calculate the proportion of the number of matches between the predicted and actual values.

$$\text{Precision} = \frac{\text{Correct Prediction}}{\text{Prediction Count}}$$

No information rate (NIR) means the precision obtained when predicting a result by selecting an arbitrary value without given data. It can be interpreted as the minimum of precision that a model can have, and the less the difference between precision and NIR, the more meaningless the model is. NIR has a value from 0 to 1,

Table 2 Six levels of Kappa coefficient

Kappa coefficient	Strength of agreement
<0.00	Poor agreement
0.01–0.20	Slight agreement
0.21–0.40	Fair agreement
0.41–0.60	Moderate agreement
0.61–0.80	Substantial agreement
0.81–1.00	Almost perfect agreement

and the closer NIR is to 1, it means that the result is consistent with the actual value even if the result is predicted as an arbitrary value. Therefore, higher NIR values indicate that the model is meaningless.

The Kappa statistic is the ratio of the actual value to the predicted value [6]. The Kappa statistic has a value from −1 to 1, meaning that 0 has no relation between the actual value and the predicted value. In the case of −1, it means that the predicted value is opposite to the actual value, so it is rare that the Kappa statistic is negative. The closer to 1, the higher the precision value can be interpreted as not being an accidental result. According to Landis and Koch's interpretation, the Kappa statistic class is divided into six levels of agreement between −1 and 1 (Table 2).

3.3 Neural Network Algorithm for Unknown Analysis

As shown in Fig. 1, a predictive model of ignition factors was made by applying a neural network algorithm. The results were output in nine types, and the result was made into a multi-label classification model.

For the data used as input, the columns with the highest precision were selected through an experiment among about 10 columns. In addition, columns that gave some improvement to the precision, but did not have a significant effect on the precision, were excluded to reduce the amount of calculation.

As the depth of the neural network gets deeper, more accurate prediction is possible, but the computational amount increases and the efficiency of the model decreases. Also, in general, as the depth of the neural network increases, the range of improvement decreases, or rather the performance decreases. For this reason, it is necessary to select a value with the highest accuracy or an appropriate value of precision compared to the amount of computation suitable for the model, and the most suitable depth of the neural network was selected based on the precision. The experiment was performed while increasing the depth of the neural network model from 1, and the depth representing the highest performance was used. The results of the experiment at each depth used the average value repeated 10 times.

When the depth of the neural network was 4, it showed the highest value with a precision of 79.18%, and thereafter, the performance tended to decrease as the depth of the neural network increased. Due to the depth of the neural network, overfitting

Input Data   **Output Data**

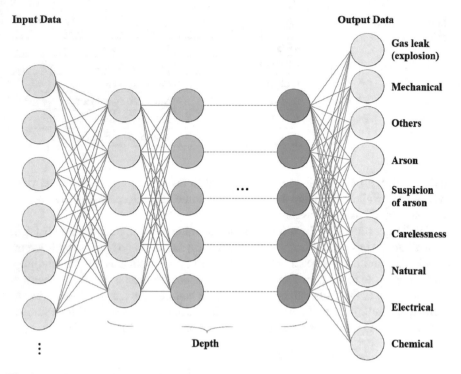

Fig. 1 Artificial neural network for analyzing fire data

occurred in the model during training, and it is inferred that precision was lowered in the actual test. When the depth of the neural network was 4, the mean value of the Kappa statistic was 0.6446, corresponding to the 0.6–0.8 divisions, which means a high level of concordance. The NIR output a value of 0.4963% (Figs. 2, 3 and 4; Table 3).

As a result of this experiment, the precision was about 80%. However, carelessness and electrical factors accounted for the most factors, and these two data accounted for 83.26% of the total. Therefore, the NIR value was intended to prove the reliability of the precision, but further research is considered necessary. In addition, items with a small number of data, such as other or natural factors, were not sufficiently trained and thus were excluded from prediction.

4 Conclusion

Data were analyzed using artificial intelligence algorithms to predict ignition factors. Since the data on one fire occurrence was recorded from various perspectives, the data was processed and used for classification. The precision of the model analyzed with artificial intelligence was 79.18%, but the result is a slightly low numerical value

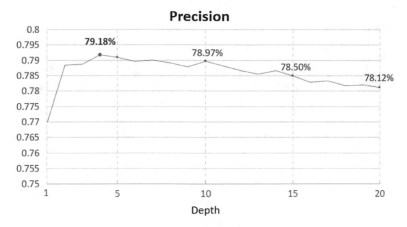

Fig. 2 Precision according to neural network depth

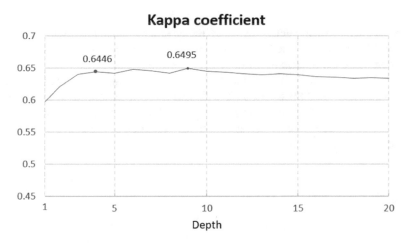

Fig. 3 Kappa coefficient according to neural network depth

to determine the ignition factor. However, it is expected to help overcome individual differences between fire inspectors and assist them in their work by analyzing ignition factors through artificial intelligence models. In addition, if the system is improved with artificial intelligence in fire inspection, it is expected that the analysis results of artificial intelligence will be provided in real time by inputting information at the site.

As a limitation of the study, since the classification model is a multi-label classification, it was difficult to make a model, and accordingly, it is considered that additional verification of precision or reliability is required. Since the structure of the fire data currently used is a database built to record all information about a fire, there is a high possibility that the results for the same fire will vary depending to the

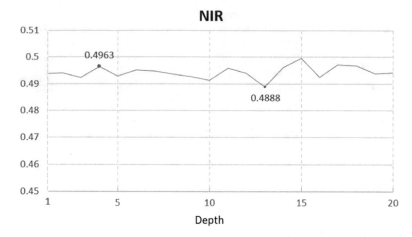

Fig. 4 NIR according to neural network depth

evaluator. Therefore, in order to use the collected data for analysis and prediction, the structure of the database needs to be improved. This will greatly affect the analysis and prediction performance of fires.

Further research will confirm the reliability of the results by using additional performance indexes for precision. It will also use longer-term data to verify the suitability of artificial intelligence algorithms. In addition, we will study with various artificial intelligence models to find a model suitable for fires and improve the performance of prediction results.

Table 3 Types of ignition factors and confusion matrix

Prediction	Actual								
	Gas leak (explosion)	Mechanical	Others	Arson	Suspicion of arson	Carelessness	Natural	Electrical	Chemical
Gas leak (explosion)	**39**	10	2	1	0	25	0	2	5
Mechanical	0	**241**	6	0	1	128	1	54	10
Others	0	0	**0**	0	0	0	0	0	0
Arson	0	0	0	**4**	0	3	0	0	0
Suspicion of arson	0	0	0	0	**0**	0	0	1	0
Carelessness	7	315	59	80	80	**3470**	31	273	120
Natural	0	0	0	0	0	0	**0**	0	0
Electrical	0	203	9	1	3	85	16	**2228**	5
Chemical	0	2	0	0	0	1	0	0	**0**

References

1. Bae GH, Jung YH, Yoo HH (2015) A study on the trend analysis regarding the fire occurrence of Jinju City. Korean Soc Geosp Inf Sci 238–241
2. Seo MS, Yoo HH (2019) Spatial econometrics analysis of fire occurrence according to type of facilities. J Korean Soc Surv Geodesy Photogram Cartogr 37(3):129–141
3. Seo MS, Yoo HH (2020) Significance analysis of facility fires though spatial econometrics assessment. J Korean Soc Surv Geodesy Photogram Cartogr 38(3):281–293
4. Kim CW, Shin DG (2020) Improved classification of fire accidents and analysis of periodicity for prediction of critical fire accidents. J Korean Inst Gas 24(1):56–65
5. Park ES, Min SH (2019) Standardization of fire factor for big data. J Korean Soc Hazard Mitig 19(4):143–149
6. Park SH, Park JH, Shin DG (2020) Verification of firefighters' heuristics through big data analysis. Korean Inst Gas 24(2):50–55
7. Richard Landis J, Koch GG (1977) The measurement of observer agreement for categorical data. Int Biomet Soc 33(1):159–174
8. Kim JS, Kim BS (2018) Analysis of fire-accident factors using big-data analysis method for construction areas. Korean Soc Civ Eng 22(5):1535–1543
9. Dutta R, Das A, Aryal J (2016) Big data integration shows Australian bush-fire frequency is increasing significantly. R Soc Open Sci 3:150241. https://doi.org/10.1098/rsos.150241
10. Lee YS, Ryu SH, Ko HA, Jeong IS (2019) A study on the standard code systematization of disaster environmental information data for earthguakes, fires and fine dusts. J Korean Geo-Environ Soc 20(12):27–32
11. Rajasekaran T, Sruthi J, Revathi S, Raveena N (2015) Forest fire prediction and alert system using big data technology. In: International conference on information engineering, management and security, pp 23–26

Development of U-Health Care Systems Using Big Data

Symphorien Karl Yoki Donzia, Haeng-Kon Kim, and Yong Pil Geum

Abstract The high mortality rate associated with cardiovascular disease requires the establishment of a personalized and ubiquitous health monitoring system. With recent advances in wireless sensor network technology, this study provides real-time data collection. Electroencephalography (EEG) is widely used for the evaluation of drowsiness, but it is not practical for careful aerial surveillance due to discomfort caused by the number of electrodes that touch the scalp. In this paper, we proposed a hearing aid-type smart sensor device connected wirelessly to a smartphone for the transmission and display of physiological data. Health care is one of the main concerns of modern people and the demand for health care systems naturally increases. We also built a big data system in this project. Building big data systems has been found to be more efficient than existing systems. This study proposes an algorithm to detect driver drowsiness through analysis of heart rate variability and compares it with EEG-based sleep scores to verify the proposed method. The ECG sensor provides various detection methods to detect RR interval data from ECG data and only transmit abnormal data. The proposed method can reduce the transmission cost and energy consumption of the sensor. We also experimentally demonstrate the energy efficiency of our method. Monitoring results using the new android app and comprehensive dysfunction experiments have been shown to improve classification accuracy.

Keywords U-Health · Care · Big data · Ubiquitous · Monitoring system · EEG

S. K. Y. Donzia · H.-K. Kim
Department of Computer Software, Daegu Catholic University, Gyeongsan-si, South Korea
e-mail: yoki90@cu.ac.kr

H.-K. Kim
e-mail: hangkon@cu.ac.kr

Y. P. Geum (✉)
Department of Innovation Start-Up and Growth, Daegu Catholic University, Gyeongsan-si, South Korea
e-mail: geum@cu.ac.kr

© The Author(s), under exclusive license to Springer Nature Switzerland AG 2021   69
H. Kim and R. Lee (eds.), *Software Engineering in IoT, Big Data, Cloud and Mobile Computing*, Studies in Computational Intelligence 930,
https://doi.org/10.1007/978-3-030-64773-5_6

1   Introduction

Motor imaging research and applications have been developed using a variety of brain wave devices (EEGs) including expensive EEG medical devices, standard EEG devices such as Emotive. Neuro Sky et al. However, recently many researchers have turned to Open BCI (not to be confused with Open BCI software).

Motor imaging research and applications have been developed using a variety of brain wave devices (EEGs) including expensive EEG medical devices, standard EEG devices such as Emotive. Neuro Sky et al. However, recently many researchers have turned to Open BCI (not to be confused with Open BCI software), an open source brain-computer interface (BCI) device. https://openbci.pl). Open BCI has its roots in crowdfunding projects. The Open BCI board is an inexpensive and versatile biosensor microcontroller that can be used to sample brain electrical activity (EEG), muscle activity (EMG), heart rate (ECG), etc. It is compatible with almost all types of electrodes and is supported by an evolving open source framework for signal processing applications (https://openbci.com) [1]. Due to its open source nature, this device offers researchers the opportunity to develop innovative BCI research and applications. This means that the software and the hardware can be modified and developed as needed. So far, only limited Open BCI research reports have been published. For example, [2] used Open BCI to control the Quadrotor [3] develop the Steady State Visual Evoked Potentials (SSVEP) framework in BCI (Brain Computer Interface) [4]. Emotion Engine is developed using Open BCI, which acts as a hub between the computer and the users. It takes the user's physiological data through body sensors and continuously estimates the user's emotional state based on the data previously collected from the user. The contribution of this study is that it could be one of the first studies to evaluate Open BCI for motor imaging applications. Brain monitoring, combined with automated EEG analysis, shortens diagnostic time and provides a clinical decision support tool that can assist clinicians in real-time monitoring applications. Electroencephalography (EEG) is used by a wide range of medical institutions to monitor and record electrical activity in the brain using electrodes placed on the scalp. EEG is essential for the diagnosis of clinical conditions such as epilepsy, depth of anesthesia, coma, encephalopathy, and brain death. Manual scanning and interpretation of the EEG takes time as these recordings can last for hours or days. It is an expensive process because it requires highly skilled professionals. Therefore, high throughput automated EEG analysis can reduce diagnostic time and improve real-time applications by marking signal sections that require further examination. Numerous methods have been developed over the years, including digital time–frequency signal processing techniques, wavelet analysis, simulated leakage integration, and multivariate techniques based on firing neurons, nonlinear dynamic EEG analysis and expert systems to be tested. It mimics the analysis of the autoregressive spectrum of the human observer and the EEG of the scalp. U-Health Care IT Convergence Technology is the amalgamation of these ubiquitous information technologies with the healthcare industry. "Health provided by enabling the collection, processing, delivery and management of information relating to human health without limitation of time and space. It can be defined as the technology necessary for the U-Healthcare service, which is a management and medical service" [5]. In existing U-Health, each sensor exists independently, so it is not possible to collect and process on-board sensors in real time. You need a U-service Embedded healthcare that can measure and process in real time. Therefore, there is a need to develop smart clothing that can be made to be suitable for attachment to clothing by miniaturizing a sensing device that patients and the general public can use without any hassle, and miniaturizing by combining modules treatment and detection communication. Currently, the wearable device market is dominated by wristband and watch type products, but various types of products are expected to increase in the future. The trend is expected to shift from accessory type products, such as bracelets and watches, to wearable devices that integrate with fabric/clothing in the future, and clothing type products are currently not large in size, but development and launch. Recent related products have been active. It seems to be driving. To build a U-shaped sanitary system, there is a need to develop a smartphone gateway module capable of software processing, such as error recovery, even if real-time acquisition, signal processing and storage in a database for a certain period of time and temporary access to the server in wireless and mobile communication situations. There is also a need to develop a medical server module that automatically analyzes and diagnoses the patient's condition, consisting of an expert rule-based and fact-based inference engine as well as situational awareness signals and information on the patient.

In Fig. 1. Automation The accessibility of "big data", was introduced for the premiere of the context of the visualization of data in 1997 [6], presenting an exception and ambitious pour la recherche bio medical with a particular accent sur

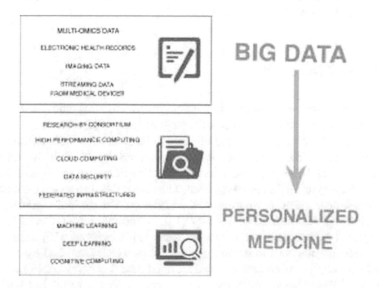

Fig. 1 Technological changes in healthcare

personalized medicine [6]. Typical biomedical portraits of Big Data are characterized by heterogeneous, multispectral, incomplete and inexact observations [7]. For
this reason, the intensive analysis of data needs ad hoc training for the modeling and
representation of complex data, the optimization of algorithms and the computing
power. In the medical field, for example, the systems are defined by the five unique
characteristics: the identification schemes of the treatments, the analysis of the data
not structured, the decision and the prediction and the traceability [8]. As a result, the
data collection is developing more quickly than the treatment and the data analysis,
which has been found in evidence of the grand card among the rapid technological
progress in the data collection and the functional nature of the biomedical information lens [9]. In this sense, the integration of the phenotypic information of the
patient card in the electronic medical dossiers (DSE) with the molecular information,
such as multi-omix data, becomes three important. If we tested a system combining
MI-based BCI and a nerve stimulator for patients with 3 stroke, combining moving
images, bar feedback and actual hand movements. In all sessions, the patient had
to imagine 120 movements with the left hand and 120 with the right hand. Visual
feedback was provided in the form of an on-screen extension bar. The FES has been
activated in tests where the correct imagination has been classified. To induce the
heat of the hand. All patients achieved high control accuracy and showed improvement in motor function. In subsequent studies, Cho et al. [8] present the results of
two patients who underwent BCI training with the first human avatar return. After
the study, both patients reported better mobility (Fig. 2).

Functional and both improved on the Fugl-Meyer Upper Limb Rating Scale. Even
though the number of patients presented in these two studies is small, they support
the idea that this type of system may provide additional benefits to the outcome of
the rehabilitation process in stroke patients.

2 Related Literature

There are several studies in the literature, including: Smart electronic noses Different
machine learning approaches to predict pathogens from VOC data. Jamal et al. We
present an electronic nose for the automatic identification of volatile chemicals and
discuss the application of the electronic nose in a variety of fields, including the
medical field [11]. Dinov suggested the classification of human pathogenic bacteria
by early detection using the electronic nose. The author then processed the data from
the electronic nose. (E-nose) How to use PCA (Principal Component Analysis) statistics. Their study demonstrated the ability of the electronic nose (E nose) to detect
early bacterial infections in the human stomach [7]. Liao et al. [10] used various
machine learning methods to predict pneumonia in VOCs identified by enosis. In
this study, we applied an ensemble-based artificial neural network (ANN) and SVM
method to VOCs collected from the e-nose sensor network to predict ventilator-
related pneumonia, and concluded that the prediction of The ANN model achieved
higher accuracy compared to SVM. Built. Another research group has proposed the

Fig. 2 TRAVEE system architecture

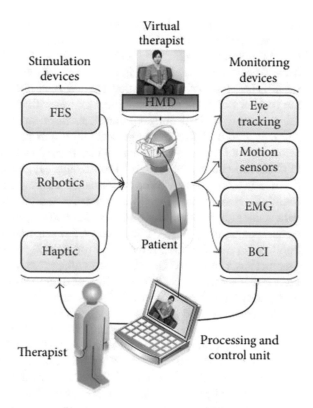

construction of an electronic code system based on neural networks (e-nose) based on a learning vector quantization method (LVQ) [12]. Because gas detection is affected by gas concentration, the researchers used a normalization method to reduce data fluctuations and reorganize measurement data based on the concentration level [12]. In this paper, we have described how to apply machine learning methods to electronic nose data independent of electronic nose data without applying feature selection. A machine learning method has been applied to predict the olfactory characteristics of VOCs [13]. In this study, a random forest (RF) algorithm was applied to compare the results with a partial least squares (PLS) regression model. The most relevant predictors were extracted from various VOCs using PCA (principal component analysis) as a feature reduction method. One of the main goals in epilepsy research is to better identify biomarkers in both acute changes and disease progression. Using neuro engineering techniques, the main goal of the laboratory is to characterize new biomarkers that indicate the timing and location of seizures, which are mainly seen in EEG signals. The most promising biomarker at the moment is the High Frequency Oscillation (HFO). HFO has a clear relationship with the epilepsy network, but how to implement it in a clinically significant and safe way is not yet clear. The main goal of this project is to develop and translate tools that enable physicians to use HFO and improve clinical care [14]. I have not found any research on seizure detection in EEG data using Spark or MapReduce, but there are studies using MapReduce to process and store EEG data. In a study, the author implements a parallel version of the Ensemble Empirical Mode Decomposition (EEMD) algorithm using MapReduce. They claim that EEMD is an innovative technology for processing neural signals, but it is both computationally and data intensive. The results show that the parallel EEMD performs significantly better than normal EEMD and confirms the scalability of Hadoop MapReduce. Another study is Hadoop and Distributed storage based on HBase for large-scale multidimensional EEG. Yahoo! Check the throughput and latency performance characteristics of Cloud Serving Benchmark (YCSB) and Hadoop and HBase. Their results suggest that these technologies hold promise in terms of latency and performance. However, at the time of the study, they found that Hadoop and HBase were not sufficiently mature in terms of stability. For data mining, recording, processing, and modeling, event-related continuous electroencephalography (EEG) is important. Choosing the right analysis tool: MATLAB Toolbox Named EEGLAB provides a programming environment and an interactive graphical user interface (GUI) for access, display and access Measurement, manipulation and storage of electrophysiological data; Allows multiple display modes of a single exam and average data [15, 16] Parametric and non-parametric tests can be performed. (Paired t-test, unpaired t-test, ANOVA) Measure. Pair/interrupt data samples. The paired/unpaired data samples were extended for more cases from two samples. The resampling method allows statistics to be made. Inference without assuming a known material probability distribution. Instead, the bootstrap method is to draw at random. Sub-sample followed by the randomization method (data mix) Sample). Signal processing and progress of the latest information. The theory has seen advancements in blind source separation. A method that attempts to find a coordinate frame. There is minimal overlap in the projection of the data. Second, this concept is ICA is a family of linear blind source separation methods. The mathematical concept of ICA is to minimize mutual information. Data projection. The ICA applies in various fields. Biosignal processing issues including: (i) Speech performance in noise separation, (ii) breakdown of functions Resonance imaging data and (iii) activity separation from brain region Mixed artifacts in wave activity cerebral. When Do a lot of statistical inference Multiple comparisons (Bonferroni, Holms method, false discovery rate, Max method, Cluster method).

3   Background of the Study

3.1   Objectives of the Study

Since Health In this study, we applied a controlled big data algorithm to identify VOCs and EEGs in uploaded data without considering the method for selecting e-nose system functions. In addition to this, there are several studies in the literature that have applied various techniques for learning machines to identify pathogens and integrate the system with the e-nose system to create a stand-alone portable e-nose system. So here we have a standalone portable e-nose system with back function removal technology to select the best VOC subset for pathogen classification.

3.2 Research Survey

The first step in this study was to create an Open BCI EEG device and cap. The design of the Spider claw hat v1 (available on the Open BCI website) has been 3D printed. I used the Open BCI GUI software to verify that the electrode placement was working correctly. The BCI-FES TRAVEE subsystem consists of the FES stimulation device, BCI monitoring device and electrooculography (EG) system, Oculus Rif virtual reality viewer and a laptop computer. And the fingers (Fig. 3). The patient is seated in a wheelchair or normal chair. The FES electrodes are mounted on the extension cords of both hands, as shown in Fig. 3, and the FES software module is started to determine the FES parameters (current and synchronization). Current impulse: up, forward, down). 10, EOG electrodes and EEG headsets are fitted and correct signal acquisition is verified. Before placing the VR viewer, the therapist sits in front of the patient and explains what he will see by showing: The virtual therapist will raise his hand as shown in Fig. 3. The therapist's right hand is the patient's right hand. Large arrows appear at the top left or right of the screen. For the virtual therapist

Fig. 3 The BCI-FES TRAVEE subsystem

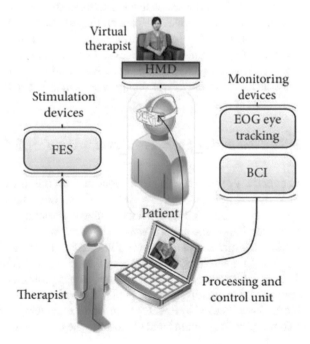

Fig. 4 The hand
rehabilitation exercise

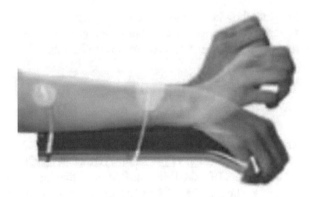

indication, the patient rings from the left or from the right. To provide VR and FES feedback, we need to create the imagined patient movement, a series of space shakes, and classifiers. First you save 4. Execute training data. Each series consisted of 20 randomized trials on the left MI and 20 on the right. We use test courses AND the signal processing algorithm presented in. Each test lasts 8 s. A clue that a beep is approaching the user in 2 s. In the second 3, the signal is displayed and marks the moment when the user should start imagining. Movement tests shown by a virtual therapist until the end. From 4.25 s while recording test data, the user sees a virtual hand marked as a signal. While moving, the nerve stimulator opens the corresponding hand of the patient. Filters and classifiers are created and 2 more of the space are saved. With VR and FES feedback, patient classification results between 4.25 and 8 s of each test are accurate. Compare all the samples and the classification results with the indices presented for each. Location, lack of achievements and more. Replace the real mirror with a virtual reality headset with the same visual feedback you need to close the patient circuit. It activates mirror neurons, but it doesn't have any drawbacks. From the mirror therapy mentioned above (Fig. 4).

3.3   Embedded Software Development Life Cycle

In general, VMware is virtual and cloud computing software. VMware is a subsidiary of Dell Technologies and is based on virtual bare metal technology. The VMware application system in Fig. 5 is not included in the study analysis, but is used to demonstrate the importance of software methodology. Analysis, design, implementation, testing and maintenance are very relevant to our research. SDLC (Software Development Life Cycle) is the process of creating or maintaining a software system. This usually involves a series of steps, from preliminary development analysis to testing and evaluating post-development software. It also consists of models and methodologies used by development teams to develop software systems, which form a framework for planning and controlling the entire development process. The SDLC

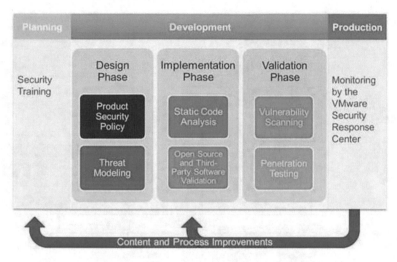

Fig. 5 Software Development Life Cycle (SDLC)

selection and adoption process is essential to maximize the possibility for organizations to successfully deliver their software, so choosing and adopting the right SDLC is a long-term management decision [17].

4 Overall Considerations

There this review describes some of the best procedures. Experimental design, visualization and description of data, or use of EEG signals for statistical analysis of inference applied in neuroscience contexts [6]. The use of EEG in the study of cortical activity in humans is a very promising scientific field because of the recent technological advances of the device, but also of the intelligence of the new pioneers of the power of artificial computation.

4.1 Overall Structure Design of U-Health Care Systems Using Big Data

The structure of the whole U-Healthcare system using smart clothing is shown in Fig. 6. It is a main board to collect and transmit biological signals by acceleration sensors, temperature sensors, blood sugar sensors and heart rate sensors, and uses Arduino Lilypad. Smartphone application software module to receive, collect and process biological signals from Bluetooth communication, and U-Health server

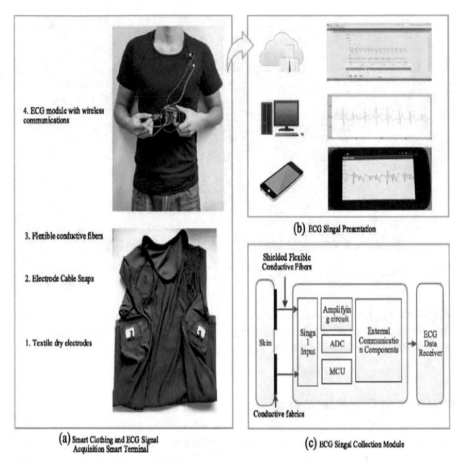

Fig. 6 The whole architecture of U-Healthcare system using smart-cloth

(medical server) to judge by storing and analyzing signals transmitted from the smartphone via the Internet on the server composed of modules.

5   Conclusion

There are previously, the biosensor device used in the U-Healthcare system for telemedicine was connected to the human body and it was difficult to operate independently. U-Health care system using smart clothing should be the most suitable smart wellness system for the health management and prevention not only of patients but also of the general public. The testing phase of these algorithms is not a complicated task since it only applies the trained model to obtain the results. Therefore, e-nose can be run independently as a stand-alone system after the configuration phase. It gives an advantage in the application Medical diagnostics, health surveillance, environmental surveillance, pathogen detection. BCI-FES system for rehabilitation after stroke. It is displayed. In addition to the stimulation device, BCI and EOG. This system of how to perform the exercise and Patient dedication and the Oculus Rif viewer Patient immersion in virtual reality. Technical performance was verified by testing the system on a healthy individual with a strong understanding of assistive technology. Healthy people had low control errors. The proportions are similar to those reported in the literature. When a U-Healthcare System is manufactured using smart clothing according to the design method shown in this document, and health conditions such as hypertensive, heart and diabetic patients should be checked from time to time. Even when it does not come, it will go a long way in creating a smart wellness system that can remotely monitor health issues and get medical help. In addition, when a medical big data system is put in place in the future, the real-time detection data and contextual data from the U-Healthcare system using smart clothing proposed in this document will be of great help in building a medical system prevention the literature.

Acknowledgements "This work was supported by research grants from Daegu Catholic University in 2020".

References

1. OpenBCI (n.d.) http://openbci.com. Accessed: 14 Aug 2015
2. Azocar AF (2014) Evaluation of the OpenBCI neural interface for controlling a quadrotor simulation, computer science. Published 2014 Corpus ID: 54208297
3. Suryotrisongko H, Samopa F (2015) Evaluating OpenBCI spiderclaw V1 headwear's electrodes placements for brain-computer interface (BCI) motor imagery application. Institut Teknologi Sepuluh Nopember, Department of Information Systems, Kampus ITS Sukolilo Surabaya 60111, Indonesia, The Third Information Systems International Conference, Available online at https://www.sciencedirect.com
4. Shamsi M (2016) Emotion recognition in Persian speech using machine learning methods. In: 3th International conference of science and engineering, Publisher. https://www.researchgate.net/publication/318040816_Emotion_Recognition_in_Persian_Speech_Using_Machine_Learning_Methods
5. Cho BH (2016) Design of U-healthcare system based on smart-cloth, Department of Information and Communication Technology at Catholic Kwandong University. J Korea Inst Inf. Electron. Commun Technol 9(2):237–242. 2005–081X(pISSN)/2288–9302(eISSN)
6. Cox M, Ellsworth D (1997) Application-controlled demand paging for out-of-core visualization IEEE Vis (1997), pp 235-244
7. Berger B, Peng J, Singh M (2013) Computational solutions for omics data. Nat Rev Genet 14:333–346
8. Cho W, Heilinger A, Xu R et al (2017) Hemiparetic stroke rehabilitation using avatar and electrical stimulation based on non-invasive brain computer interface. Int J Phys Med Rehabil 05(04)
9. TRAVEE (2017) Virtual terapist with augmented feedback for neuromotor recovery. https://travee.upb.ro/. Last visit Sept 2017. https://sites.google.com/a/umich.edu/staceylab/home/human_eeg

10. Liao Y-H, Wang Z-C, Zhang F-G, Abbod MF, Shih C-H, Shieh J-S (2019) Machine learning methods applied to predict ventilator associated pneumonia with pseudomonas aeruginosa infection via sensor array of electronic nose in intensive care unit. Sensors 19(8):1866 [Online]. Available: https://www.mdpi.com/1424-8220/19/8/1866

11. Villiere A (2018) Random forests: a machine learning methodology to highlight the volatile organic compounds involved in olfactory perception. Food Qual Prefer 68:135–145

12. Ahmed L, Edlund A, Laure E (2016) Parallel real time seizure detection in large EEG. In: IoTBD 2016 - International Conference on Internet of Things and Big Data, Department of Computational Science and Technology, Royal Institute of Technology, Stockholm, Sweden

13. Makeig S, Debener S, Onton J (2004) Mining event-related brain dynamics, University of California, San Diego. Trends Cogn Sci 8(5):204–10. This publication at https://www.researchgate.net/publication/8584104, https://doi.org/10.1016/j.tics.2004.03.008, Source PubMed

14. Poboroniuc MS, Ortner R, Allison BZ, Guger C (2017) Preliminary results of testing a BCI-controlled FES system for post-stroke rehabilitation. In: Proceedings of the 7 Graz brain-computer interface conference 2017, Graz, Austria, 18–22 Sept 2017

15. Muzafar T, Wadhwa RK, Diganta B, Kothari NSY (2013) Evaluation of mirror therapy for upper limb rehabilitation in stroke. IJPMR 24(3):63–69

16. Poongodi T, Sanjeevikumar P Internet of Things (IoT) and E-Healthcare System – A short review on challenges. Galgotias University. See discussions, stats, and author profiles for this publication at https://www.researchgate.net/publication/331876656

17. Donzia SKY (2019) Development of predator prevention using software drone for farm security. Thesis, Daegu Catholic University, Department of Computer Software, p 22

Analyses on Psychological Steady State Variations Caused by Auditory Stimulation

Jeong-Hoon Shin

Abstract Contemporary society is seriously threatened with food as part of the world due to the continuous increase in world population, the degradation and decline of agricultural lands due to high industrialization, climate change and the aging of the population. Therefore, modern society is studying different solutions to solve human food. In this paper, a framework for precision agriculture using IoT Gateway is proposed for solving human food, and the productivity of crops must be increased first. IoT solution through architecture, platforms and IoT standards, or the use of interoperable IoT technologies beyond the adopters in particular, simplifying existing proposals. Connecting different sensors, connected devices, developing intelligent breeding systems as much as possible. One of our aims is to manage and challenges. We provide a techniques and technologies applications during our work. The result shows that the advantages of various types of sensors for agriculture services in their decision making. And a proposed architecture for Agriculture Mobile services based on Sensor Cloud substructure that helps farms and IoT applications are effective in intelligent farming system. In this study, therefore, we have presented a technique for applying the stimuli allowing all subjects to maintain their psychologically steady state by using the customized stimuli optimized for individual users, and have attempted to promote the psychologically steady state of the subjects by using stimuli tailored to each user. For that reason, we performed quantitative analyses of brainwaves in the users by using the auditory stimuli and to investigate the variations in the psychologically steady state of the users based on a two-dimensional analytical model on emotions.

Keywords Brainwave · Neuro-feedback · Auditory stimulus · Psychological steady state

J.-H. Shin (✉)
Daegu Catholic University, Gyeongsan-si, South Korea
e-mail: only4you@cu.ac.kr

© The Author(s), under exclusive license to Springer Nature Switzerland AG 2021   81
H. Kim and R. Lee (eds.), *Software Engineering in IoT, Big Data, Cloud and Mobile Computing*, Studies in Computational Intelligence 930,
https://doi.org/10.1007/978-3-030-64773-5_7

1 Introduction

Various studies related to brainwaves have been conducted, and in particular, there is a heightened interest in areas associated with control of the brain activation state of subjects, such as neuro-feedback treatment and training. Neuro-feedback represents a treatment and training technique that provides users with the capability to measure and visualize their brainwaves in real time and improve their cognitive abilities [1]. The neuro-feedback therapy and training technique is non-invasive, and has found wide-ranging applications in medicine, healthcare and many other fields due to the advantage that it is safer and less demanding than other therapy and training techniques. Moreover, neuro-feedback therapy and training may also serve as an effective and alternative treatment for patients with cognitive and mental disorders [2].

This neuro-feedback therapy and training allows users to visualize and control their brainwaves. However, many users experience difficulty with this when they engage in neuro-feedback for the first time. Furthermore, there is a problem in that the effectiveness varies, depending on the users, when the same external stimuli are provided [3, 4].

Thus, the selection of stimuli is important in the neuro-feedback therapy and training, and those stimuli show differences, depending on the physical and psychological characteristics of individual users. In this study, therefore, the stimuli optimized for users based on auditory stimuli were selected and presented.

The difference in the effectiveness of neuro-feedback treatment and training, which varies depending on individuals, can be determined through brainwave measurement and analysis of its characteristics. Brainwaves refer to the electrical signal generated from the cerebral nervous system [4]. Brainwaves represent an aggregation of nerve cell membrane potentials, the size of which indicate the extent of activity among nerve cells [5]. Brainwaves are classified into delta, theta, alpha, beta, gamma, and SMR waves. Additionally, brainwaves occurring in many different situations, which reflect various states of humans, while engaged in such things as sleep, meditation, language, numeracy, cognition, and concentration, show differences.

In this study, therefore, we performed EEG analyses of subjects according to the changes in auditory stimuli by using single and complex tone auditory sound stimuli. Through that, the performance of single and complex tones was verified, and the changes in psychological steady states were analyzed through the two-dimensional analysis model of emotion. If the stimuli optimized for each subject, based on the analyzed data are selected and provided, it is expected that the effectiveness of neuro-feedback therapy and training can be maximized, and the subjects will be able to control their brainwaves more efficiently.

2 Experimental Design

2.1 Auditory Stimulus Sound Source

A total of 25 critical bands with acoustically sensitive characteristics in the audible frequency band, the range of audible frequencies for humans, were selected to control the brainwaves of subjects at a specific point of time more efficiently, and the sound sources, each lasting 8 s, were created by using the critical band and the center frequency of the critical band [6]. For the single tone, the sound in the center frequency of the critical band (50–19,000 Hz) was created as the sound source. For the complex tone, 3 types of sound sources were created. Type 1 was based on the center frequency (Q factor 40) while Type 2 was based on the critical band (Q factor 8). Finally, Type 3 was created as a complex tone combining Type 1 and Type 2. The Q factor conversion formula is presented in Fig. 1. Figure 2 shows an example of the auditory stimulus sound source frequency of the complex tone, created on the basis of the formula.

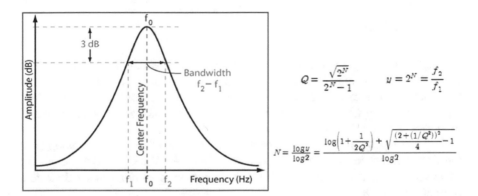

$$Q = \frac{\sqrt{2^N}}{2^N - 1} \qquad \upsilon = 2^N = \frac{f_2}{f_1}$$

$$N = \frac{\log \upsilon}{\log 2} = \frac{\log\left(1 + \frac{1}{2Q^2}\right) + \sqrt{\frac{(2 + (1/Q^2))^2}{4} - 1}}{\log 2}$$

Fig. 1 Q factor calculation formula

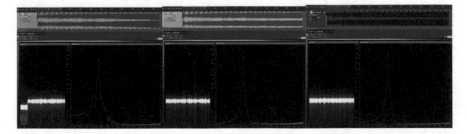

Fig. 2 250 Hz sound source of the complex tone (Type 1, Type 2, and Type 3)

2.2 Experimental Design

The experiment was carried out in an independent space blocking external noise, and inaccessible by outsiders, in order to collect stable and reliable data. A total of 56 subjects were enrolled in the experiment, which consisted of 28 males and 28 females. The subjects were healthy adults between the ages of 20–29.

Prior to the start of the experiment, personal information, including the presence of hearing loss, use of hearing aids, and history of auditory organ surgery of the subjects was examined. Then, the audible frequency band of the subjects was measured. After the measurement was completed, preparations were made for experiments in a relaxed state. Subjects ready for the experiments were measured, starting from background brainwaves. The brainwaves of the subjects were measured while they were in a relaxed state, with their eyes closed for 30 s. Then, the auditory stimulus experiment was performed with 1 set of auditory stimulus sound sources of single tones and 3 sets of auditory stimulus sound sources of complex tones.

2.3 Data Analysis

To analyze the changes in the psychological steady state, a two-dimensional analysis model (z-score) of Valence and Arousal emotions was used [7]. Valence measures the emotions of pleasantness and displeasure, which is related to prefrontal lobe alpha waves. Arousal evaluates the extent of relaxation and excitement, also related to prefrontal lobe alpha and beta waves.

In the analysis, this two-dimensional model was used for the area ranging from the first quadrant to the fourth quadrant by using both Valence and Arousal. The two-dimensional model used to analyze the changes in psychological steady state is presented in Fig. 3.

3 Results of Experiment

3.1 Quantitative Brainwave Analysis

In this study, quantitative brainwave analysis was performed through the fast Fourier transform of EEG signals collected from subjects. Analyses examined the relative energy variations in delta, theta, alpha, beta, and gamma waves over time, to determine the differences between background brainwaves, the brainwaves caused by auditory stimuli, and the difference between single and complex tone auditory stimuli, and furthermore, statistical analyses based on those findings were conducted.

Fig. 3 Two-dimensional
model for analyzing the
changes in psychological
steady state

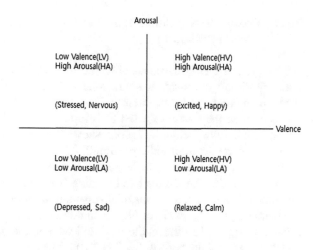

3.1.1 **Comparison of Non-stimulus State and Auditory Stimulus State**

The brainwaves of subjects reacted to the single and multiple tone auditory stimuli, resulting in a reduction of delta and gamma waves. Conversely, reactions to the same stimuli led to an increase in alpha waves. That is considered attributable to the fact that the data from the subjects were collected while they were in a relaxed state, with their eyes closed, during the measurement of their background brainwaves. Figure 4 shows the variations in each band of frequency generated in response to auditory stimulation compared to background brainwaves.

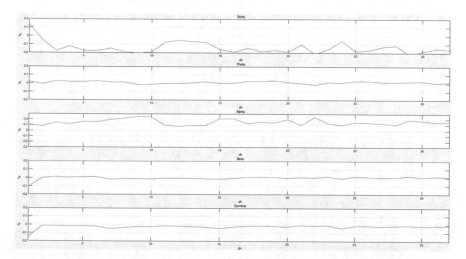

Fig. 4 Variations of energy by band, caused by auditory stimulus sound source stimulation (center frequency of 2510 Hz, single tone sound source) (band from top)

3.1.2   Comparison of Single Tone Auditory Stimuli and Complex Tone
Auditory Stimuli

An increase of theta waves was observed in the subjects exposed to complex tone auditory stimuli, compared to single tone auditory stimuli. Theta waves are associated with memory and concentration [8]. This result shows that subjects were more focused when they were exposed to complex tone auditory stimuli, compared to single tone auditory stimuli. Figure 5 shows the increase in theta waves, caused by single and complex tone auditory stimuli.

In particular, the delta wave decrease was relatively higher in response to auditory stimuli of single tones in the band ranging between 2160 and 2925 Hz, while the decrease in the delta waves was relatively smaller in response to auditory stimuli in the band range of 5850 Hz or higher. In addition, the increase in alpha waves was relatively larger when the auditory stimuli were in the band ranging between 1600 and 4850 Hz, and the band of 8600 and 10,750 Hz.

An increase of alpha waves was observed in subjects exposed to complex tones, as opposed to single tones. Alpha waves show a leisurely and calm state [9]. This suggests that subjects felt more relaxed and comfortable when they were exposed to auditory stimuli of complex tones than auditory stimuli of single tones. Figure 6 shows an increase in alpha waves, caused by auditory stimuli of single and complex tones.

Beta waves of channels 22, 23, 31, and 32 showed an increase in the brainwaves of subjects exposed to auditory stimuli of complex tones, compared to single tones. Beta waves are observed in a woken state, and are the most dominant and powerful of the brainwaves [10]. The beta wave increase in channel 31 and 32, which shows the parietal lobe, indicates sensory activation. This suggests that subjects reacted more sensitively to external auditory stimuli in this region of the brain when they

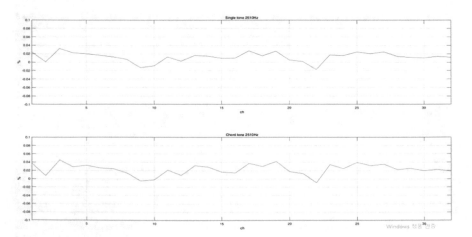

Fig. 5 Variations of theta wave energy, caused by auditory stimulus sound source stimulation (center frequency of 2510 Hz, sound sources of single and complex tones)

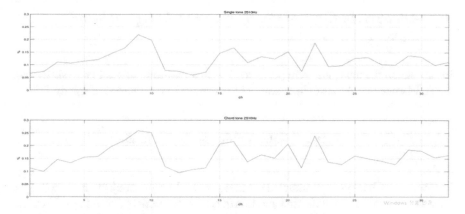

Fig. 6 Variations of alpha wave energy, caused by auditory stimulus sound source stimulation (center frequency of 2510 Hz, sound sources of single and complex tones)

were exposed to complex tones, compared to single tones, which resulted in sensory activation. In addition, there was a tendency that other activities became desensitized when the beta waves were increased in this same area. Figure 7 shows an increase in beta waves, caused by auditory stimuli of single and complex tones.

In particular, the reduction of delta waves was relatively larger in the subjects exposed to auditory stimuli of complex tones in the band ranging between 250 and 8600 Hz, while the increase of theta waves was relatively larger in the subjects exposed to complex tones in the band ranging between 350 and 13,750 Hz. For channels 31 and 32, the reduction of delta waves was relatively larger when an auditory stimulus of complex tones was provided, compared to single tones.

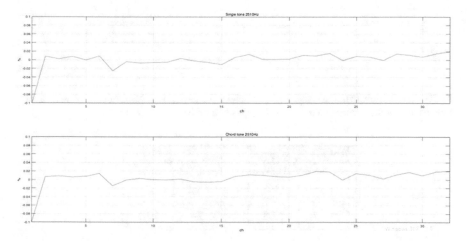

Fig. 7 Variations of beta wave energy, caused by auditory stimulus sound source stimulation (center frequency of 2510 Hz, sound sources of single and complex tones)

3.2   Analysis Analyses of Changes in Psychological Steady State

In this study, the analyses were performed with a two-dimensional analysis model, using Valence and Arousal, to examine the changes in psychological steady state.

3.2.1   Comparison Between Non-stimulus State and Auditory Stimulus State

In the case of background brainwaves measured while the subjects had their eyes closed, the distribution in the first and fourth quadrants, which showed a positive and comfortable state of emotion, comprised 49.8% of the entire distribution as shown in Figs. 8 and 9. When the auditory stimuli were provided, the distribution in the first and fourth quadrants, again indicative of positive emotions, accounted for 48.1% of the entire distribution. The reduction by 1.7% is considered attributable to the effect of external auditory stimuli, compared to the background brainwaves.

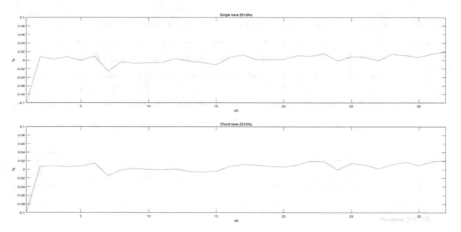

Fig. 8 Distribution based on analyses of changes in psychological steady state with a two-dimensional model (non-stimulus state—25 types of auditory stimuli X 4 set)

Non-stimulation		Auditory stimulation	
18.8 ₂	16 ₁	28.2 ₂	19 ₁
31.4 ₃	33.8 ₄	23.6 ₃	29.1 ₄

Fig. 9 Ratio of distributions based on analyses of changes in psychological steady state with a two-dimensional model (non-stimulus state—25 types of auditory stimuli X 4 sets)

3.2.2   Comparison Between Auditory Stimuli of Single and Complex Tones

As shown in Figs. 10 and 11, the distribution in the first and fourth quadrants, indicative of a positive state of emotion, comprised 44% of entire distribution in the experiment with the auditory stimuli of single tones. Meanwhile, in the experiment conducted using complex tones, the distribution in the first and fourth quadrants, indicative of positive states of emotion, comprised 51.5%, 46.5%, 50.5%, respectively, which was higher than when single tone auditory stimuli were used. This suggests that complex tones induced subjects to feel a state of happiness, comfort, and calmness, when compared to single tone stimuli.

In addition, the increase in the distribution was the largest in the first and fourth quadrants when the auditory stimuli of complex tone type 1, created based on the center frequency, was provided. By contrast, the increase in the distribution was the smallest, as opposed to other auditory stimulus sound sources of complex tones, in the first and fourth quadrants when complex tone type 2, created based on the

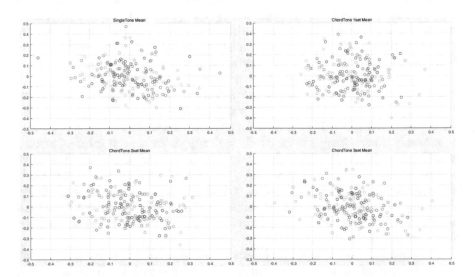

Fig. 10 Distribution based on analyses of changes in psychological steady state with a two-dimensional model (single tones, complex tones type 1, type 2, and type 3)

Single tone		Complex tone type 1		Complex tone type 2		Complex tone type 3	
33 ₂	16 ₁	26 ₂	21.5 ₁	28.5 ₂	18.5 ₁	25.5 ₂	20 ₁
23 ₃	28 ₄	22.5 ₃	28 ₄	25 ₃	28 ₄	24 ₃	30.5 ₄

Fig. 11 Ratio of distribution based on analyses of changes in psychological steady state with a two-dimensional model (single tones, complex tones type 1, type 2, and type 3)

critical band, was introduced. This suggests that complex tones, produced in the key frequency, made subjects feel the greatest comfort.

4 Conclusion

In this study, EEG tests of subjects were analyzed by using auditory stimuli, and based on that, statistical analyses were implemented, along with investigations on the changes in psychological steady state by using a two-dimensional analytical model. When external auditory stimuli were provided to subjects, a decrease in both delta and gamma waves, and an increase in alpha waves was observed, compared to the background brainwaves occurring in a relaxed state with eyes closed. This is considered to be significantly associated with the opening and closing of the eyes and an influx of external stimuli. The auditory stimuli of complex tones led to an increase of both theta and alpha waves in most channels, including the increase of beta waves in channels 22, 23, 31, and 32, as opposed to the auditory stimuli of single tones. This suggests that complex tones led subjects to be more focused and feel more relaxed and comfortable, and that the subjects reacted more sensitively to external auditory stimuli in channels 22, 23, 31, and 32, relative to single tone auditory stimuli. As beta waves were increased near the parietal lobe corresponding to channels 31 and 32, it was considered that complex tones were involved in sensory activation.

Moreover, complex tone auditory stimuli led to a relatively greater reduction of delta waves in channels 31 and 32, compared to single tones. This suggests that an increase in beta waves led to a relative decrease in delta waves.

We analyzed the changes in the psychologically steady state of subjects, and the results showed that the rate of distribution in the first and fourth quadrants, indicative of a positive state of emotion, increased by 7.5%, 2.5%, and 6.5%, respectively, in the experiment with auditory stimuli of complex tones, compared to single tones. This suggests that complex tones led subjects to feel more comfortable, stable and more focused, in contrast to single tones, which is consistent with the results of quantitative analyses.

Future studies will investigate the indicators serving as the basis for classifying the emotional states of users. In addition, studies will be conducted on various stimulation techniques that can help make subjects feel more mentally balanced. The results of those studies are expected to provide techniques optimized for each user to increase their psychological stability.

Acknowledgements This work was supported by research grants from Daegu Catholic University in 2020.

References

1. Kong M, Kim JS (2019) The study of neurofeedback effects on cognitive abilities of adults with intellectual disability. J Rehabil Welfare Eng Assist Technol 13(1):50–56
2. Lecomte G, Juhel J (2011) The effects of neurofeedback training on memory performance in elderly subjects. Psychology 2(8):846–852
3. Kim JS, Lee MK (2017) The effect on concentration and BDNF according to visual cognition training using neuro-trophic. J Korean Soc Living Environ Syst 24(5):648–653
4. Lee K, Jung YJ (2017) Design of high frequency boosting circuits compensating for hearing loss. J Inst Electron Inf Eng 54(3):138–144
5. Yoon JH, Cho SH (2017) The EEG analysis of first impression including facial shape and color. J Curr Instruct Stud 10(2):15–27
6. Lee CH, Hong SK (2018) A study on the hearing characteristic-based equalizer design for the elderly. J Digit Contents Soc 19(4):779–787
7. Ha JM, Park SB (2017) Assessment of color effect on the indoor color schemes and illuminance change—focused on prefrontal EEG alpha and beta signal analysis. J Arch Inst Korea 33(10):57–65
8. Byun YE (2017) Effect of prefrontal lobe neurofeedback training for reducing adolescent theta waves. J Korea Acad Ind Coop Soc 18(12):459–465
9. Shim JM, Kim CS, Goo BO (2008) The effects of a-wave music and art appreciation on hand function. J Korean Soc Phys Therapy 20(1):75–79
10. Choi SJ (2019) Beta-wave correlation analysis model based on unsupervised machine learning. J Digit Converg 17(3):221–226

Flipped Learning and Unplugged Activities for Data Structure and Algorithm Class

Jeong Ah Kim

Abstract In this research, new blended teaching method was suggested for algorithm learning since students have difficulties to understand algorithms and to make program in programming languages. Programming from the scratch has been very common teaching method but it has requires too much efforts to learn the syntax which is not the essential subject in perspectives of computational thinking. Video lecture about explaining the algorithm in code and running example is provided for pre-class activities of flipped learning. So, students can understand the basic concepts by code and running examples. Unplugged activities are designed to evaluate how much the students studied in pre-class and identify the part which students cannot understands. For post-class, just homework for identifying the application of algorithm in real life is issued. I evaluate the effectiveness for 2 groups: experimental group with suggested method and other group with just flipped learning and compare the outcomes with test of computational thinking assessment. As the result, experimental group showed higher outcomes and improvement of self-efficacy is little higher.

Keywords Computational thinking · Algorithm education · Flipped learning · Unplugged learning

1 Introduction

Computational thinking, as coined by Jeannette Wing, is a fundamental skill for all to be able to live in today's world, a mode of thought that goes well beyond computing and provides a framework for reasoning about problems and methods of their solution [1].

J. A. Kim (✉)
Department of Computer Education, Catholic Kwandong University, Gangneung 25601, South Korea
e-mail: clara@cku.ac.kr

© The Author(s), under exclusive license to Springer Nature Switzerland AG 2021
H. Kim and R. Lee (eds.), *Software Engineering in IoT, Big Data, Cloud and Mobile Computing*, Studies in Computational Intelligence 930,
https://doi.org/10.1007/978-3-030-64773-5_8

The ability to write and understand computer programs has become an essential skill for engineers to learn but it is not easy way. To teach the programming in the class, IT teacher should care the students individually since each student's performance could be different. This is why we used flipped learning for programming education since student study the basic syntax and concept in pre-class and identify the insufficient points for understanding. In-class, student practice the problem with teacher and teacher can give feedback and more information to each student personally. For flipped learning, teacher should design the course carefully.

This research was designed for identify the effectiveness of unplugged activities for in-class phase of flipped learning. For evaluate the effectiveness of unplugged activities for in-class phase, experimental research was designed for 2 year with same grade and same course enrollment students. At first year, algorithm class was designed with flipped learning with video materials for pre-class, algorithm representation practices for in-class, and code analysis for post-class. At second year, instead of algorithm representation practice, unplugged activities for understanding algorithm was replaced. To get more valid and stable evaluation result, "Bebra Task" contest evaluation was used for measurement tools.

In Sect. 2, we describe the background knowledge of computational thinking, Algorithmic thinking, Unplugged activities, and flipped learning. We briefly introduce the related previous works in Sect. 3 and explain our research methods in Sect. 4. Research results are given in Sect. 5.

2   Background

2.1   *Computational Thinking*

According to Wing [2], computational thinking is a fundamental skill for everyone, not just for computer scientists. To reading, writing, and arithmetic, we should add computational thinking to structure every child's analytical ability. Also, Wing defined that computational thinking involves solving problems, designing systems, and understanding human behavior, by drawing on the concepts fundamental to computer science. Computational thinking is the thought processes involved in formulating problems and their solutions so that the solutions are represented in a form that can be effectively carried out by an information-processing agent [3]. Although coming from computer science, computational thinking is not only the study of computer science, though computers play an essential role in the design of problems' solutions. It is a very important and useful mode of thinking in almost all disciplines and school subjects as an insight into what can and cannot be computed [1]. Computational thinking is the process of approaching a problem in a systematic manner and creating and expressing a solution such that it can be carried out by a computer.

2.2 Algorithm Education

Algorithmic thinking is the ability to think in terms of such algorithms as a way of solving problems. This approach automates the problem-solving process by creating a series of systematic logical steps. An algorithm is the starting point for writing a computer program. Algorithm is a part of computational thinking since the four cornerstones of computational thinking are decomposition, pattern, abstraction and algorithms.

Algorithm education using Pico board with Scratch and flowchart improve the computational thinking ability [4]. Scratch programming increase the computational thinking through STEAM education and Scratch with physical computing curriculum increase the computational thinking ability [5, 6]. In other words, programming is very good way to increase the computational thinking for problem solving. The Algorithm is a fundamental concept in computer science teaching [7]. Various approaches have been proposed to support teaching of algorithms, involving use of graphical and verbal representations of algorithms [8]. The Algorithm is a fundamental concept for teaching Computer Science in Secondary Education. There are graphic representations of algorithms that can be used for teaching the algorithm to middle and high school students. Many university students were using visual programming language to create computational competency and problem-solving capabilities.

2.3 Flipped Learning

Typical class has been focused on teacher's activities. These days, new challenges have come to teachers. Role of teacher is changing, from lecturer to facilitators. It means teacher should give more opportunities for students to engage the learning. The role of teacher is changing in smart and active learning methodologies [9]. Now teacher is as a facilitator in learning. Teaching and learning are being modified due to innovations in education.

Flipped classroom as an educational technique that consists of two parts: interactive group learning activities inside the classroom, and direct computer-based individual instruction outside the classroom [10]. In flipped learning, student try to understand the content at home and teach at class. It means students explain their understanding to teacher or other students to assess their pre-class efforts [11]. Feedbacks or adjustments individually to each student at in-class lead to increase the understanding of course materials [12]. Integrating the pre-class and in-class make the students to have more responsibilities so that students participate the class more actively and the outcomes can be increased as well as self-efficacy [13]. Doing "homework" in class gives teachers better insight into student difficulties and learning styles [14, 15]

2.4   Unplugged Activities

Unplugged coding aims to teach programming concepts through the use of games or activities that can be done offline using tangible objects, such as paper and markers. Offline coding is a good way to engage younger students without the use of technology. It helps students better understand computer science concepts through role-playing, analogies and other visual exercises [16].

3   Related Previous Works

Avancena et al. [17] presents the initial stage of developing an algorithm learning tool for the students of the Information Systems course. Suggested tool applies the concept of Algorithm Visualization (AV) technology and was used as an aid for learning basic algorithms such as searching and sorting. Lee [18] showed the efficiency of flipped learning of algorithm course at computer engineering department. This research was design with online class and small task at class with team discussion. del Olmo-Munoz [19] evaluated whether the inclusion of the so-called unplugged activities favors the development of CT in the students of the early years of Primary Education. In this research 3 questions were evaluated: the development of their CT skills, their motivation towards the proposed instruction, and the influence of students' gender in the two previous areas. The intervention was designed on a selection of activities extracted from Code.org courses, and was divided into two phases, one in which one group worked with unplugged activities and the other with plugged-in activities, and a second where both groups worked with plugged-in activities. Özyurt and Özyurt [20] designed flipped classroom approach for 94 students in the scope of Introduction to Programing and Algorithm course during 14 weeks and produced the result which majority of the students expressed positive views about the flipped classroom approach. Jang [21] proposed the flipped learning model consisting of five sets that combine the flipped learning and practice to improve student motivation and self-directed learning. Also, this paper analyzes the learning effect by applying it to the algorithm lecture of computer engineering and presents problem and utilization plan according to the result.

4   Research Design and Methodology

This study aims to identify the efficiency and effectiveness of unplugged activities for in-class activities when flipped learning technology was adapted for algorithm course.

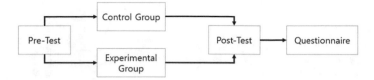

Fig. 1 Design of the research

Control Group: Session Structure		
Phase	Time	Teaching & Learning
Pre-Class	25 ~ 30	Video Class
		On-Line Material
In-Class	10 ~ 20	Offline Quiz
	20 ~ 30	Teaching subjects
	20 ~ 30	Algorithm Representation with Flowchart and NS Chart
	15 ~ 25	Team Activities
Post-Class	20 ~ 40	Practice with Code
		Interaction and Reflection

Experimental Group: Session Structure		
Phase	Time	Teaching & Learning
Pre-Class	25 ~ 30	Video Class
		On-Line Material
In-Class	10 ~ 20	Offline Quiz
	20 ~ 30	Teaching subjects
	20 ~ 30	Unplugged Activity
	15 ~ 25	Team Activities
Post-Class	20 ~ 40	Practice with Code
		Interaction and Reflection

Fig. 2 Design of session

4.1 Design of Research

Based on the objectives of this study, a quasi-experimental design with control and experimental groups was proposed. Regarding the development of the experience, it was structured in two main training phases of instruction interspersed with pre- and post-tests. Figure 1 describes the process implemented in the development of the research.

This study was designed for 2 years from 2018 to 2019. The sample of the research consisted of a total of 52 students of the 2nd year of Catholic Kwandong University. Almost these students did not take a programming course before university and take a C programming course at first year of university. Control group and experimental group were formed by 26 students. For control group, course was designed with only flipped learning. For experimental group, course was designed with flipped learning and unplugged activities. Figure 2 describes the session structure implemented in the development of the research.

4.2 Design of Course

For this study, 5 subjects were selected for data structure and algorithm course. (1) Tree and tree traversal algorithm, (2) Graph and graph traversal algorithm, (3) Sorting algorithm, (4) Search algorithm (5) Shortest Path problem.

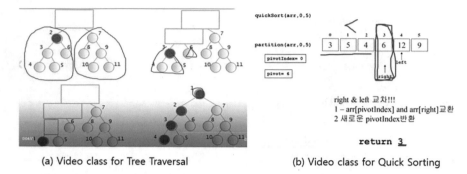

(a) Video class for Tree Traversal   (b) Video class for Quick Sorting

Fig. 3 Sample of pre-class (flipped learning)

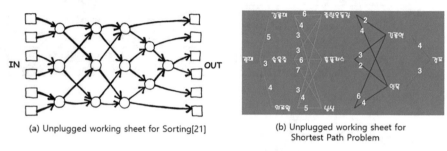

(a) Unplugged working sheet for Sorting[21]   (b) Unplugged working sheet for
                                                  Shortest Path Problem

Fig. 4 Sample of unplugged working sheet (in-class activities)

For each subject, on-line video classes were provided. These video classes were made by research group. There are many useful resources for pre-class such as MOOC, YouTube, and another online tutoring system. But, in this study, these resources are suggested for further reading and learning by students and course material was developed by lecturer itself (Fig. 3).

For experimental group, unplugged coding sheets were provided and team activities are performed. There are many unplugged working sheets what are developed by University of Caterbury [22]. In this study, development of unplugged working sheet is not the goal so that we applied many unplugged working sheet from [22]. Even though, some working sheet was developed shown in Fig. 4.

With these working sheet, student performed team discussion of algorithm and solve the real problem applying the algorithm shown in Fig. 5.

4.3 Design of Evaluation

For evaluate the learning outcome, each year mid-term and final-term test were performed and compare the results between 2 groups. And student's satisfaction was analyzed. These instrument may produce the reliable measurement but might

Fig. 5 Unplugged activity: in-class

not mean a valid results because of the lack of consensus in the field in terms of CT definition and lack of measurement tools of CT. For more valid and reliable assessment, many researches applied Bebras learning model [23–25]

The Bebras Tasks are a set of activities designed within the context of the Bebras International Contest, a competition born in Lithuania in 2003 which aims to promote the interest and excellence of primary and secondary students around the world in the field of Computer Science from a CT perspective. Each year, the contest launches a set of Bebras Tasks, whose overall approach is the resolution of 'real-life' and significant problems, through the transfer and projection of the students' CT [26]. For all these features, the Bebras Tasks have been pointed out to more than likely be an embryo for a future PISA (Programme for International Student Assessment) test in the field of Computer Science [27].

In this study, the instrument consisted of an initial selection of 8 items from the category Group 6(High school) from 2014 to 2017 edition of 'International Bebras Contest' (Fig. 6).

5 Research Results and Conclusion

Considering the objectives of this study, results were obtained for both groups in each year (Pre-and Post-test) for each of the proposed areas to be compared in a grouped way.

Regarding learning outcome, Table 1 shows the average score of midterm and final examination which scale was from 0 to 10.

The descriptive data indicated that there were more achievement of unplugged and flipped learning blending group than flipped learning group even though there were slight difference in blending group (midterm test 8.699) in comparison to the flipped learning group (midterm test 8.739) on scale Data set (F-test = 0.48284) is fit for comparison. After testing the assumptions, t-test for pre-test and post-test of computational thinking for problem solving was analyzed. According to these results, there were some meaningful differences between pre-and post-test results of the students' computational thinking for problem solving [t(26) = −2.31, p(0.022657) < 0.05]. In detail, the understanding and analysis capability was increased more than other subjects.

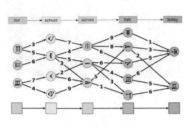

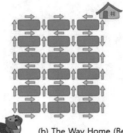

(a) Best Translation (Bebras Challenge 2014)
Concepts - Abstraction (AB), Evaluation (EV),
Generalization (GE)

(b) The Way Home (Bebras Challenge 2017)
CT Skills - Algorithmic Thinking (AL),
Decomposition (DE), Evaluation (EV)
CS Domain - Algorithms and programming
Tags - Route, Backward searching, Black holes

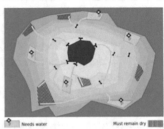

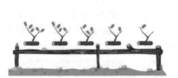

(c) Sorting Brach (Bebras Challenge 2017)
CT Skills - Abstraction (AB), Algorithmic Thinking (AL)
CS Domain - Algorithms and programming
Tags - Sort, Swap, Memory

(d) Beaver Gates (Bebras Challenge 2016)
Skills - Abstraction (AB), Evaluation (EV)

Fig. 6 Sample of Instrument for evaluation from Bebras challenge

Table 1 Comparison of Learning Outcome

	Mid team	Final team	Improvement	Ratio of improvement students (%)
A group (2018)	8.739	9.347	0.608	88
B group (2019)	8.699	9.493	0.794	96

Regarding student's satisfaction, there was no significant difference since size of group was not big and participant belong to department of computer education so that their motivation to experience with learning technology. But the results of self-efficacy of blending group are significantly higher than flipped class.

Regarding student's evaluation of Bebras Tasks, students experienced unplugged activities got the higher scores (Table 2).

Table 2 Comparison of achievement of Bebras challenges

	Mean	Std. dev.	N	t	p
FL + unplugged	4.65	4.006	26	−1.789	0.04285
FL	4.19	3.441	26		

In this research, we tried to explore the effects of unplugged activities as in-class performing for flipped learning of algorithm courses. The quantitative part was conducted in control group and experimental group for 2 years. The scores of computational problem solving ability were measured through the assessment tools.

As a result, unplugged activities help students to understand algorithm and apply it to solve real problem since there are no obstacle to learn the syntax of programming languages. With control group, algorithm representation activities with flowchart and NS chart were used for in-class of flipped learning. But algorithm representation might be another programming task not problem solving process. In this research, understanding and applying should be prior to writing a program.

The most unplugged working sheets used in this study came from [21] so that some working sheet is not proper to university students. Since unplugged activities can be applied to all ages, it is necessary to develop more higher level working sheets for algorithm classes.

References

1. Sysło MM (2015) From algorithmic to computational thinking: on the way for computing for all students. In: ITICSE'15: innovation and technology in computer science education conference 2015. p 1
2. Wing J (2006) Computational thinking. Commun ACM 49(3):33–36
3. Wing J (2008) Computational thinking and thinking about computing. Philos Trans R Soc Math Phys Eng Sci 366(1881):3717–3725
4. Shi S-B (2015) The improvement effectiveness of computational thinking through scratch education. J Korea Soc Comput Inf 20(11):191–197
5. Kim TH (2015) Steam education program based on programming to improve computational thinking ability. Ph.D thesis. Department of Computer Education, Jeju National University
6. Kim JH (2016) Development and application of EPL and physical computing curriculum in elementary schools for computational thinking. Master thesis. Educational Graduate School. ChungJu National Educational University
7. Tucker AB, Bernat AP, Bradley WJ, Cupper RD, Scragg GW (1995) Fundamentals of computing I. McGraw Hill, New York
8. Scanlan D (1988) Should short, relatively complex algorithms be taught using both graphical and verbal methods, six replications. ACM
9. Prakash J (2016) Teachers role as facilitator in learning. Sch Res J HumIty Sci & Engl Lang 3(17)
10. Bishop JL, Verleger MA (2013) The flipped classroom: a survey of the research. In the proceeding of 2013 ASEE annual conference and exposition, vol 30, 9. pp 1–18
11. Kim JA, Heo HJ, Lee HeeHyun (2015) Effectiveness of flipped learning in project management class. Int J Softw Eng Appl 9:41–46
12. Bergmann J, Sams A (2012) Flip your classroom: reach every student in every class every day. In: Bergmann J, Sams A (eds) International society for technology in education
13. Namik K, Boae C, Jeong-Im C (2014) A case study of motivation and self-efficacy. Educ Technol 30(3):467–492
14. Fulton KP (2012) 10 Reasons to flip. New styles of instruction 94(2):20–24
15. O'Flaherty J, Phillips C (2015) The use of flipped classrooms in higher education: a scoping review. Int High Educ 25:85–95
16. Bell T, Fellows M, Witten I (2008) Computer science unplugged: Off-line activities and games for all ages. https://www.csunplugged.org/en/

17. Avancena AT, Nishihara A, Kondo C (2015) Developing an algorithm learning tool for high school introductory computer science, vol 2015. Hindawi Publishing Corporation Education Research International
18. Lee S-H (2016) A case study of flipped learning in algorithm class. Proc Winter Conf Korean Comput Inf Soc 24(1):175–178
19. del Olmo-Munoz J, Cozar-Gutierrez R, Gonzalez-Calero JA (2020) Computational thinking through unplugged activities in early years of primary education. Comput Edu 150:103832
20. Özyurt O, Özyurt H (2017) A qualitative study about enriching programming and algorithm teaching with flipped classroom approach. Pegem Eğitim ve Öğretim Dergisi 7(2):189–210
21. Jang S (2017) Design of effective teaching-learning method in algorithm theory subject using flipped learning. J. Korea Inst Inf Commun Eng 21(5):1042–1048
22. https://csunplugged.org/en/
23. Jung U, Lee Y (2017) The applicability and related issues of Bebras challenge in informatics education. J Korean Assoc Comput Educ 20(5):1–4
24. Roman-Gonzalez M, Moreno-Leon J, Robles G (2019) Combining assessment tools for a comprehensive evaluation of computational thinking interventions. Computational thinking education, Springer, pp 78–99
25. Tang X, Yin Y, Lin Q, Hadad R, Zhai X (2020) Assessing computational thinking: a systematic review of empirical studies. Comput Educ 148:103798
26. Roman-Gonzalez M, Moreno-Leon J, Robles G (2017) Complementary tools for computational thinking assessment. Computational thinking education, Springer, pp 154–159
27. Hubwieser P, Mühling AM (2014) Playing PISA with Bebras. In: WiPSCE'14, Proceedings of the 9th Workshop in primary and secondary computing education. pp 128–129

Study on Agent Architecture for Context-Aware Services

Symphorien Karl Yoki Donzia and Haeng-Kon Kim

Abstract With the evolution of the web services paradigm, there is a need for an effective mechanism to offer a better service experience to users in a dynamic adaptation process. In particular, the context knowledge can be adapted to dynamic adaptation taking into account the context of the user and the context of the device. Creating context-sensitive services today is a complex task because there is no adequate infrastructure support in a ubiquitous IT environment. This paper we proposed a service-oriented contextual architecture for creating fast contextual services. The objective of this study is to develop an intelligent agent platform that applies the developed intelligent agent platform to the field of medical applications. The results describe the general process of a situational awareness system and the design considerations for the situational awareness system. In addition, it summarizes the characteristics of the existing situational awareness system and presents a trend analysis of the system. In conclusion, we propose future work for a better situational awareness system.

Keywords Software architecture · Context-Aware service · Web service · Ubiquitous

1 Introduction

This framework is an architecture for providing optimal services in a ubiquitous environment that adapts to the user's situation, and can automatically detect user behavior and the service model provided to the user, thus becoming the basis of the provision of intelligent services.. It makes sense as a study. The proposed frame- work has the ability to include an Event-Condition-Action (ECA) rule-based triggering system that can automatically respond to events based on the user context collected by the

S. K. Y. Donzia · H.-K. Kim (✉)
Department of Computer Software, Daegu Catholic University, Gyeongsan-Si, South Korea
e-mail: hangkon@cu.ac.kr

S. K. Y. Donzia
e-mail: yoki90@cu.ac.kr

© The Author(s), under exclusive license to Springer Nature Switzerland AG 2021
H. Kim and R. Lee (eds.), *Software Engineering in IoT, Big Data, Cloud and Mobile Computing*, Studies in Computational Intelligence 930,
https://doi.org/10.1007/978-3-030-64773-5_9

sensor. The concept of ubiquitous computing is a basic technology that can immediately provide the information or services needed by users by functionally and spatially connecting all existing objects in real space, including humans [1, 2]. In a ubiquitous environment, every computer with functions such as collecting, processing and communicating information is functionally and spatially connected, and a method is needed to immediately provide the necessary information or services to users [3]. In the previous study, the creation of a context ontology [4] was carried out taking into account the user's situation and data mining techniques [5, 6] using the ontology. In this article, we describe the behavior of users according to their situation. The modern supply chain is becoming increasingly global in the number of companies involved, and it is moving from a closed and static environment to an open and dynamic one. In such an environment, the ability to respond to changes in demand in a timely and cost effective manner is important, and this can be at the heart of a company's economic strength. Personal relationships were once. However, more flexible and dynamic skills and techniques are needed to establish a supply chain that responds to changes in demand, but there is insufficient empirical research on this. Ubiquitous technologies such as USN and wired/wireless communications networks Transportation, environment, health and wellness, crime prevention and infrastructure-based defense Ubiquitous services in the areas of finance, culture and tourism, education and administration.. This increases the comfort of life and improves the quality of life. Guarantee of security and improvement of the well-being of the citizens thanks to the systematic management of the city, a new mountain IT at the cutting edge of technology which innovates all the functions of the city, such as the creation of companies. In recent years, both domestic and foreign countries have developed the characteristics and advantages of cities. Focus where possible, strengthen the competitiveness of cities and pursue the realization of u-City with the common goal of seeking synergies. There is [7]. However, in the flood of information that occurs in the ubiquitous environment, it is very difficult to distinguish the necessary information from users. Converted. In particular, mobile devices such as laptops, cell phones and PDAs. As energy increases, the proper distribution of information is nothing less than information gathering. Effort required. In such an environment, the contextual system helps users to select information and When you provide a service, you maximize the usefulness of the information produced. In other words, instead of allowing users to find the information they are looking for, select and provide information that is appropriate for the user's situation. Es. Personalization of u-City applications and services means compute and communication Smart objects with application capabilities Cognition and adaptability (context sensitive) are required. Use of contextual information Almost all the information available at the time of the self-interaction. This is usually the location, identification, activity and status of a person, group or object. Etc. At this point, the status information is the same as the user's current activity. It can be personal and the same as the device you are currently using. It can be technical, but also the temperature, location or time. It could also be environmental. The situation is also the physical location. Or it can be a primitive, like a zip code, and One or more fundamental circumstances, such as the name or address of the water, can also be composed of a compound [8]. Contextual service personalization is a way to collect and exchange contextual information. Through the process of recognition through the ring, interpretation and reasoning, users personalized services adapted to the situation. Emergency For example, when a fire occurs in zone A, the house fire sensor sends a disaster emergency request to the u-City Integrated Operations Center. To serve. The integrated operations center contacts the relevant zone A fire department, fire trucks have been dispatched and the inhabitants of zone A are informed of the fire and the information is transmitted to the u-Traffic system. city and by the u-Traffic service, to divert vehicles passing through Zone B where traffic conditions are mild. This allows fire trucks to enter quickly. By observing the above scenario, the situational awareness system is activated in zone A. Recognize the fire situation and notify residents of the fire, The results of inferences such as securing a road d 'access for the rapid deployment of fire trucks. And connect each service to deliver. It is the fire truck safe fast access road, driver fast detour road Provide adequate services to users, such as information services. The reasoning is based on a proposition that you already know. It is defined as the derivation of a new proposition. Yes so If there is no reasoning in the situation or if it is a principled reasoning After identifying the sulfur, only the fire alarm went off or only the fire truck was hurried. As a result, the emergency disaster service would have ended. The advantage of reasoning is the feet. Proactively derive and provide adaptation methods for real events Complexity, connectivity. One of the main challenges of intelligent e-commerce is to develop a needs identification system. It is a personalized notification system to meet the needs of users in the e-commerce market. Our proposed notification system sets it apart from traditional marketing information systems or supply chain management systems. In that the recall system supports a more personalized individual approach, such as one-on-one interaction with users to help users find specific products. Most current reporting systems only provide organized and routine information, such as a daily schedule. Information. In addition, most notification systems lack flexibility. Indeed, today there is no built-in user-centric needs identification mechanism that can automatically adapt to changing conditions. Developing smart spaces can be difficult and expensive without the proper support of IT architecture, which is seen as the first big challenge. Over the years, a significant number of intelligent space architectures, middleware, and frameworks have been developed using various technologies. However, there is still no generic model to guide the design of the system. Furthermore, the main drawback of these systems is their weakness in supporting situational awareness, which is a key feature of these spaces and aids in adaptive service work tremendously. Some proposed architectures focus on one aspect (e.g., the physical layer) and ignore other important aspects (e.g., the adaptation of context services), limiting usability to specific cases of smart spaces [9].

In the Fig. 1. Automation The results of all models are shown in Fig. 2a, b for micro-F1 and macro-F1, respectively. Each graph contains separate F1s for intents and spaces. Looking from left to right, the performance of the contextual and non-contextual test suites are shown separately. You can also use contextual models to improve the performance of contextual test suites without compromising the performance of contextual test suites. For all contextual models, the number above the bar graph indicates the relative rate of change from the model to the non-contextual

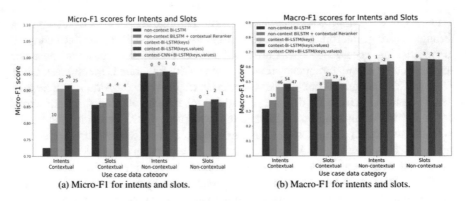

(a) Micro-F1 for intents and slots.   (b) Macro-F1 for intents and slots.

Fig. 1 Numbers on bar graphs show relative % improvements compared to non-contextual baseline

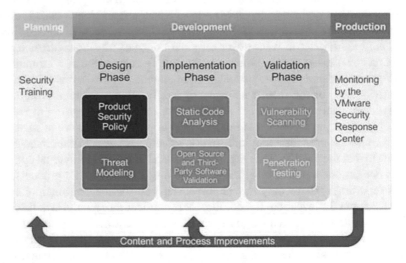

Fig. 2 Software development life cycle (SDLC)

BiLSTM model. Positive changes indicate better performance [10]. Context BiLSTM achieves higher performance in context use cases compared to a two-step strategy using rule-based reorganizations. Our model can learn more complex contextual behaviors that are directly optimized. You can compare it to the hand-designed re-sorting logic. It's also easy to extend modeling with new functionality by adding new contextual use cases to your data, but rule-based systems require more manual maintenance. I was able to slightly increase the performance of the reorganization by optimizing the rule-separated weights, but as the contextual functionality increased, these strategies were out of reach because these strategies were not scalable. The rest of this paper is organized as follows: Sect. 2 presents some Related Literature work on the current adoption of for Context-Aware Services. Section 3 will introduce our Background of the Study. Section 4 explains how Overall Architecture of

Context-Aware Services can help drive digital efficiency. Finally, we conclude in Sect. 5.

2 Related Literature

In recent years, research on context-sensitive computing has moved towards scalable and reconfigurable architectures. It can be used to automatically recognize user needs in the context of an application and adjust system functions and interaction patterns accordingly [11]. One challenge that designers of context-sensitive applications face is the lack of generic infrastructure to develop context-sensitive applications. In addition, existing applications have mainly focused on location information. Several researchers have proposed and prototyped a generic architecture to support contextual applications. Schilit infrastructure Ubiquitous computing [12] is probably in this direction. It is agent based and supports meetings. Context for devices and users. The application is Architecture uses this contextual information Other agents such as Device Agent, User Agent and Active Agent Orientation to provide users with proactive services. Dey's Context Toolkit [6] aims to facilitate the development of context-sensitive applications by providing a small set of generic "base" classes from which application developers can derive specific classes. These classes are organized around high-level application capabilities that take context into account: collecting sensor data, combining data from multiple sensors, and converting sensor data into alternative formats. The Contextual Toolkit provides a standardized way of working on the syntactic part of a contextual system. However, inferences about contextual information must be implemented by domain. And the application. This is an architecture which is the implementation phase, but not the execution phase. Middle where [13] is location sensitive middleware that separates applications from location sensing technology. It incorporates multiple location technologies to give your application a unified view of the locations of moving objects and can be added as these technologies become available. Probabilistic reasoning techniques are used to resolve conflicts and infer a specific location. Other sensor data. Environment Where also allows the Mobile objects applications and their environment. The SOCAM (Service-Oriented Context-Sensitive Middleware) project [14] is a rapid development and prototyping of contextual mobile services. It supports the acquisition, interpretation, discovery and dissemination of context, and its main feature is to support context inference, which can derive high-level implicit context from a context explicit low level. The context is represented by a predicate written in OWL (Web Ontology Language) which offers flexibility. The interpreter collects the context data from the distributed context providers and presents it to the client in a predominantly processed form. Gaia [15] aims to make physical space intelligent by providing services for the management of space and its associated state. The context provider is a data source for contextual information. Other agents can poll or listen to an event stream that continues to send context events. You can also publish the context that you provide with a context provider search service that allows other agents to discover it. The context history service can store and query past contexts. The software architecture of a computer program or system is the structure of a system that includes software components, the externally visible properties of these components and the relationships between them [16]. With interesting architecture Middleware systems for smart spaces have been developed over the years. Some of them were layered architectures and the most widely used agent technology like that proposed in [15, 17]. Smart-M3 was based on the slate architecture model [18]. Some proposed architectures were pure agent systems (multi-agents each) like [9–19]. Fan et al. [20] have proposed a web services architecture for the home. Some of the proposed architectures were pure sensor network architectures, such as those proposed in [21, 22]. Cook [23] studied some intelligent multi-agent space systems to provide insight into the role of multi-agent research in the context of intelligent environments. They concluded that many intelligent environment software agents rely on expert systems or machine learning technologies. When multiple residents enter the smart environment, it becomes much more difficult to provide the services provided by the smart environment. Tracking multiple residents is clearly a difficult problem, and indeed has turned out to be NP-Hard. Researchers working in the field admit that scalability issues are overwhelming, according to the authors. Another challenge of the software architecture of the multi-agent intelligent environment is to define a lightweight, simple and scalable form. According to Nakajima [24], one way to reduce development costs is to reuse existing software whenever possible. These abstractions are useful for building your surrounding intelligence without considering low-level details. He described the four middleware infrastructures developed as part of the project to create an ambient intelligence environment. He confirmed that the middleware infrastructure provides a high level of abstraction for specific application areas in order to hide complexities such as implementation and context awareness. Memon et al. [25] carried out a review of the literature on the state of the art of the assisted environment. The AAL (Living) framework, systems, and platforms identify essential aspects of the AAL system and explore important issues from a design, technology, quality of service and user experience perspective. They found that most AAL systems are limited to a limited set of features and ignore many essential aspects of the AAL system. Standards and techniques are used in a limited and isolated way, while quality attributes are often not sufficiently covered.

3   Background of the Study

3.1   Objectives of the Study

The technology is moving beyond personal computers to the trend of microprocessors embedded in everyday objects and consumer electronics. The advancement of sensor networks and devices with computing capabilities has provided us with the necessary technologies. Create a smart space. Services are provided in a way that is not visible in advance. The goal of our smart space is to design the system with UML software metrology architecture diagram to provide users with more convenience, energy savings, safety and huge benefits for the elderly or disabled. living alone. It also offers a multi-agent architecture to build an intelligent lounge focused on the contextual aspect. The solution is expected to improve capture security and increase the competitiveness of auction houses by introducing a digital auction system to replace traditional analog systems, which can sometimes be unsanitary.

3.2 Research Survey

The main The proposal system has objective providing optimal services to users in a ubiquitous computing environment, it is important to consider spatiotemporal information closely related to objects and user behavior. We design a spatio-temporal ontology to consider the user's situation and propose a system structure that can actively mine the user's behavior and service patterns using this. And The proposed system is a framework for intelligently mining the user's behavior and service patterns in consideration of the user's location over time and the relationship with the object, and is based on a trigger system. The second part of our study will be aboutu-City services whichare different mainly public and private. It is divided into u-Public and u-Environment, u-Traffic, etc., and these services These are services provided under control. Private Services For u-Healthcare, u-Home, u-Work, u-Education services It is possible to be frequented by the provincial service provider. Synergy effects can be achieved when connecting with the city operations center [26]. The city's operations center is located in each of the city's communications networks, transportation networks and facilities. Receive city information from hood sensor and analyze it in an integrated way Efficiently manage and manage city and analyze city information. It is provided in real time. U-City's urban operations center is Regardless of infrastructure, space, time and terminal Provides permanent access environment, situational awareness and a reusable person Sharing a number of similar control systems through plastic construction Minimization of redundant investments thanks to the change of new requirements Towards a flexible software infrastructure architecture for All acceptance. In addition, from the integration between unitary services to the open service structure to achieve vertical integration between organizations and horizontal integration between services placed at each level. Service architecture) as the base address.

3.3 Embedded Software Development Life Cycle

In general, VMware is virtual and cloud computing software. VMware is a subsidiary of Dell Technologies and is based on virtual bare metal technology. The VMware application system in Fig. 2 is not included in the study analysis, but is used to demonstrate the importance of software methodology. Analysis, design, implementation, testing and maintenance are very relevant to our research. SDLC (Software Development Life Cycle) is the process of creating or maintaining a software system [20]. This usually involves a series of steps, from preliminary development analysis to testing and evaluating post-development software. It also consists of models and methodologies used by development teams to develop software systems, which form a framework for planning and controlling the entire development process. The SDLC selection and adoption process is essential to maximize the possibility for organizations to successfully deliver their software, so choosing and adopting the right SDLC is a long-term management decision [27].

4   Overall Architecture of Context-Aware Services

UML (Unified Modeling Language) is a standard visual model. Language used in the analysis, design and implementation of software systems. UML is a way to visualize software using a collection of diagrams. The unified modeling language UML aims to become the industry standard for software development and documentation. But the research project also UML with a lot of effort. By describing UML in a simplified way, we can say that the language provides a set of different diagrams to highlight different aspects of the software. Different diagrams are typically used differently at different stages of software development. implementation solution which are detail on Figs. 3 and 4.

4.1   Use Case Diagram

Use cases illustrate the units of functionality provided by the system. The main objective of the use case diagram is to assist in the development of applications for health in the use and connection of health professionals through the services provided by the medical, dental, pharmaceutical science laboratories and clinics. Use cases illustrate the units of functionality provided by the system. Including "user" relationships. Use case diagrams are generally used to convey the high level of functionality of our system. offers a multi-agent architecture to build an intelligent lounge focused on the contextual aspect. The solution is expected to improve capture security and increase the competitiveness of auction [27].

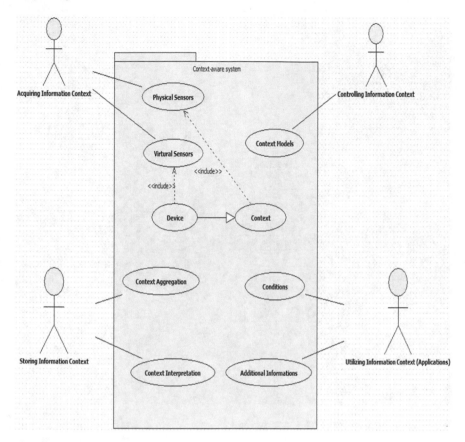

Fig. 3 Use case diagram of context-aware services

4.2 Class Diagram

The class diagram shows "Architecture of Context-Aware Services entities which are a connection of the solution is expected to improve capture security and increase the competitiveness of auction [27].

4.3 Sequence Diagram

The Sequence diagrams showing Fig. 5 detailed flows for specific use cases that facilitate the functionality of an intelligent agent platform that applies the developed intelligent agent platform to the field of medical applications.

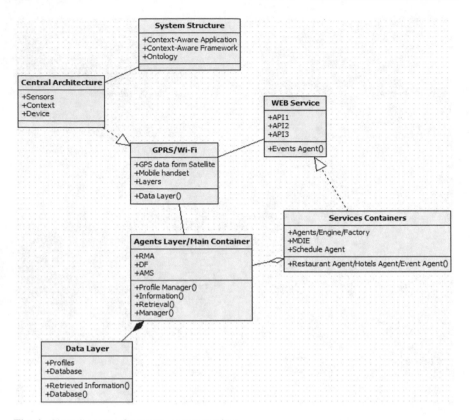

Fig. 4 Class diagram of context—aware services

4.4 Role of IoT

The Information technology that collects information in real time by connecting sensors to objects and exchanging information between individual objects on the Internet. What people don't need to adjust or estimate. Essential skills: sensor programming and network programming. The Internet of Things (IoT) is a network of physical objects containing embedded technologies that communicate, sense, or interact with the internal state or external environment. People connect people more appropriately. The process gets the right information to the right entity at the right time. Data transforms data into actionable insights, insight and wisdom. Internet and physical objects connected to each other to make better decisions. The expansion of industry, regulation and agriculture will lead to the standardization of the control and operation of drone-powered IoT. Equipped with a variety of IoT devices, it can be used to form an integrated IoT platform that operates in the air. Everything is connected or connected wirelessly. Drones provide precise information about the truth of the ground and provide more precise images as they are closer to the ground. You can use drones to adjust and measure distances from terrain, calculate depth levels, and

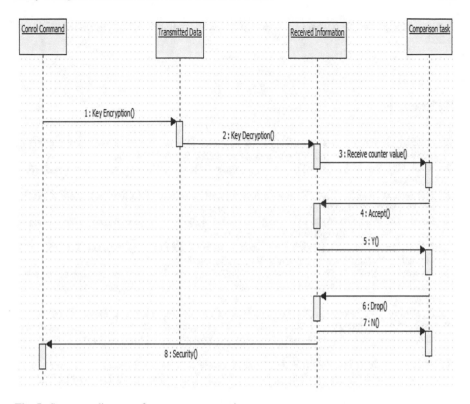

Fig. 5 Sequence diagram of context-aware service

measure many IoT applications. The Internet of Things (IoT) is a network of physical objects containing embedded technologies that communicate, sense, or interact with the internal state or external environment.

5   Conclusion

Creating context-sensitive services today is a complex task because there is no adequate infrastructure support in a pervasive IT environment. In this work, we have proposed a service-oriented contextual architecture for the creation of rapid contextual services. Efficient smart space is to design the system with UML software metrology architecture diagram to bring users more convenience, energy saving, safety and huge benefits for the elderly or disabled, which has been proposed in detail. Future work will be a continuation of this study precisely to develop a detailed implementation of the system into real-time system applications that can be integrated using the current framework.

Acknowledgements "This work was supported by research grants from Daegu Catholic University in 2020"

References

1. Anselin M, Harry 1C, Tim F (2003) An ontology for context-aware pervasive computing environments. Workshop ontologies and distributed systems, IJCAI Press
2. Khedr M, Karmouch A (2004) Negotiating context information in context-aware systems. IEEE Intell Syst
3. Strimpakou M et al (2005) Context modeling and management in ambient-aware pervasive environments. Workshop on Location and Context-aware
4. Strimpakou MA, Roussaki LG, Anagnostou ME (2004) A context ontology for pervasive prevision. National Technical University of Athens
5. Brisson L, Collard M (2008) An Ontology driven data mining process. In: The tenth international conference on enterprise information systems
6. Bellandi A, Furletti B, Grossi V, Romei A (2007) Ontology-driven association rules extraction: a case of study, the international workshop on contexts and ontologies: representation and reasoning
7. Byung-cheol L, Yong-ju L (2007) u-City business model and u-service. TTA J 112:72–82
8. Oh-Byeong K, Nam-yeon L (2007) The situation recognition system using equipment collaboration korean structural evaluation methodology. J Korea Intell Inform Syst Soc G, 13(2):27–41, 6
9. Miraoui M, El-etriby S (2016) Agent-based context-aware architecture for a smart living room 3Al-Leith Computer College, Umm Al-Qura University, KSA2 Faculty of computers and information. Menoufia Univ Int J Smart Home 10(5):39–54
10. Naik VI, Metallinou A, Goel R (2018) Context aware conversational understanding for intelligent agents with a screen. 1 Arizona State University, 2 Amazon Alexa Machine Learning
11. Anagnostopoulos C, Tsounis A, Hadjiefthymiades S (2004) Context awareness in mobile computing: a survey. proceedings of mobile and ubiquitous information access workshop, Mobile HCI '04, Glasgow, UK
12. Chen H, Finin T, Joshi A (2004) A context broker for building smart meeting rooms. In: Proceedings of the AAAI symposium on knowledge representation and ontology for autonomous systems symposium, pp 53–60
13. Ranganathan A, Al-Muhtadi J, Chetan S, Campbell R, Mickunas MD (2004) Middle where: a middleware for location awareness in ubiquitous computing applications. In: Proceedings of the 5th ACM/IFIP/USENIX international conference on Middleware, Canada, pp 397–416
14. Gu T, Pung KK, Zhang DQ (2004) A middleware for buildingcontext-aware mobile services. In: Proceedings of IEEE Vehicular Technology Conference (VTC), Milan, Italy
15. Cook DJ, Youngblood M, HeiermanEO, Gopalratnam K, Rao S, Litvin A, Khawaja F (2003) MavHome: an agent—based smart home PERCOM '03. In: Proceedings of the first IEEE international conference on pervasive computing and communications, pp 521–524
16. Bass L, Clements P, Kazman R (2003) Software architecture in practice (2nd ed.). Addison –Wesley, Reading, MA
17. Madhusudanan J, Hariharan S, Manian AS, Venkatesan VP (2014) A generic middleware model for smart home. Int J Comput Net Inf Secur 6(8):19–25
18. Honkola J, Laine H, Brown R, Tyrkkö O (2010) Smart -M3 information sharing platform, computers and communications (ISCC). In: IEEE Symposium on, Riccione, Italy, pp 1041–1046
19. Reinisch C, Kofler MJ, Kastner W (2010) ThinkHome: a smart home as digital ecosystem . In: 4th IEEE international conference on digital ecosystems and technologies (IEEE DEST), Dubai, UAE, pp 256–261

20. Fan YJ, Yin YH, Xu LD, Zeng Y, Wu F, IoT-based smart rehabilitation system
21. Familiar MS, Martınez JF, Lopez L (2012) Pervasive smart spaces and environments: a service-orientedmiddleware architecture for wireless Ad Hoc and sensor networks. In: International journal of distributed sensor networks, Hindawi Publishing Corporation, pp 1–11
22. Sun Q, Yu W, Kochurov N, Hao Q, Hu F (2013) A multi -agent -based intelligent sensor and actuator network design for smart house and home automation. J Sens Actuator Net 2:557–588
23. Cook DJ (2009) Multi -agent smart environments. J Ambient Int Smart Environ IOS Press 1:47–51
24. Nakajima T (2010) Case study of middleware infrastructure for ambient intelligence environments chapter handbook of ambient intelligence and smart environments. Springer, pp 229–256
25. Memon M, Wagner SR, Pedersen CF, Beevi FHA, Hansen FO (2014) "Ambient Assisted Living Healthcare Framework" platforms, standards, and quality attributes. Sensors 14:4312–4341
26. Jeon H (2007) Une unité clé pour la réalisation des services publics/privés u-City Alcohol. TTA J 112:46–54
27. Donzia SKY (2019) Development of predator prevention using software drone for farm security. Department of Computer Software. Daegu Catholic University Thesis 2019.022

Secure Transactions Management Using Blockchain as a Service Software for the Internet of Things

Prince Waqas Khan and Yung-Cheol Byun

Abstract The Internet of Things (IoT) has enabled communication anywhere between physical devices. Currently, concerns have been raised about suspicious transactions in IoT systems. Suspicious transactions may have a logical structure inconsistent with current information in the context of IoT. This article describes suspicious transactions in IoT systems and manages them using the blockchain as a service software plans. This study builds software-specific components for blockchain functions to implement in IoT networks. This study was conducted using Hyperledger Fabric as a blockchain service to test the software-defined blockchain components blockchain. The model was evaluated using average transaction throughput and latency. We observed that using blockchain as a software service system can provide excellent performance and security.

Keywords Internet of things (IoT) · Suspicious transactions · Blockchain · Hyperledger fabric

1 Introduction

With the rapid development of the internet of things, incumbent operators are beginning to make a big difference in digitization. In the age of the Internet of Things, millions of devices and links to the Internet of Things are a big problem for data management. Many systems today use centralized systems to manage IoT devices, creating privacy, and security concerns while also managing IoT data [1]. Because of the blockchain's level, traceability, and latency, the blockchain has generated considerable interest in the IoT region. However, integrating current blockchain technology into the IoT is difficult due to the lack of scalability and high cost. Various blockchain

P. W. Khan · Y.-C. Byun (✉)
Department of Computer Engineering, Jeju National University, Jeju-si, Republic of Korea
e-mail: ycb@jejunu.ac.kr

P. W. Khan
e-mail: princewaqas12@hotmail.com

platforms have unique advantages in IoT data management mode. In work by Jiang et al. [2], the methods to integrate multiple blockchains to manage IoT data efficiently is explained. Their solution builds a shared server access system using a single blockchain as a control number. Other blockchain platforms dedicated to the defined IoT environment serve as the basis for all IoT devices. Their model is based on a notary approach and engages transactions in this way for certainty. They analyzed the system's performance evaluated through extensive testing.

Blockchain is a crucial technology for decentralized system management and is becoming increasingly popular when implementing smart grids and healthcare systems. However, due to the high resource requirement and low scalability due to frequent and frequent transactions, mobile devices with limited resources are limited. Integrate edge accounts to offload mining operations from mobile devices to cloud resources. This integration ensures reliable access, distributed processing, and tamper-free archiving for scalable and secure transactions. Therefore, critical issues related to security, scalability, and resource management must be addressed to achieve effective integration. Nyamtiga et al. [3] draw on peer-to-peer technology and blockchain to implement the Internet of Things (IoT) design, supported by cutting-edge computing to achieve the level of security and scalability required for integration. To successfully integrate the blockchain into the IoT system, they explored the existing blockchain and related technologies to create solutions to anonymity, integrity, and resilience. Research has been conducted to verify proper architecture requirements, and some researchers have applied a combination to popularize specific applications. Despite these efforts, the anonymity, resilience, and integrity of functional and secure distributed data storage still need to be further investigated.

Figure 1 illustrates some of the benefits that blockchain and IoT can bring together. High data security, robust data validation process, etc. Using the IoT blockchain use case, companies will get a complete information security network that can provide facilities that third-party providers cannot offer. IoT blockchain use cases can provide a validation process based on consensus algorithms to make data entry fairer. Users can trust this combo for added privacy of data and privacy. A fully transparent network is a trusted network. The problem with our shared centralized server is the lack of full transparency. Most companies also need a personal channel, so the IoT blockchain use case can only provide a particular chain between two parties.

Furthermore, IoT devices can use this channel to communicate with each other without problems. The main components of the IoT are sensors and RFID tags [4]. These tags allow anyone to track the object and ensure full reliability.

The rest of the article comprises related work explained in Sect. 2, Proposed architecture is explained in Sect. 3. Chapter 4 describes the transaction flow. Part 5 contained the simulation setup and evaluation results. We concluded the article in the conclusion section.

Fig. 1 Blockchain and IoT:
Possible use cases

2 Related Work

Internet of Things and Blockchain technology is dominating many areas of research. The Internet of Things can find good results in various fields, but the BBC offers a reliable communication system for home communications. IoT devices for data exchange have been implemented since the blockchain. This led to a step-by-step integration process. The biggest downside to this combination is that blockchain's internal processing of information is complete and fast. On the other hand, due to many IoT devices, current blockchain solutions generate data at a faster rate than they can afford. On the other hand, Blockchain files cannot run on IoT devices due to a lack of resources. This way, the two technologies won't merge into their current state. The work by the Biswas et al. [5] uses a network of a local peer network to address these issues and provide solutions to close the gap. Domestic and global peer-to-peer data connections offer the number of transactions sent globally without connecting to the complete information on the local server. The test values show a significant reduction in the weight and size of the reader through international comparison.

Blockchain is a distributed ledger that contains transactions related to blockchain. The ability to create, store, and send digital assets in a decentralized, decentralized, and tamper-proof manner is of great practical value for IoT systems. The biggest challenge in providing blockchain as a service to the Internet of Things is the host environment. Peripherals are usually very limited in terms of computer resources and available bandwidth, and the cloud or fog can be a potential host. The author evaluates the use of fog and cloud as possible platforms [6]. Recently, the Software-Defined Industrial Internet of Things (SDIIoT) has emerged. This is considered an effective way to manage the Internet of Things dynamically. SDIIoT implements multiple SDNs to improve scalability and flexibility, forming a physical distributed control plane that processes large amounts of data generated by industrial equipment. However, it is difficult to reach consensus among many SDN controllers as the core of multiple SDNs. With the advent of blockchain technology, some IoT network management functions could be transferred from centralized systems to distributed certification bodies. The cloud-based blockchain application has been successfully implemented. However, the Internet of Things does not require the specific functionality of the blockchain.

An example of this is a competitive consensus. In their study, Samaniego et al. [7] explored the blockchain's capabilities in editing and hosting software-defined components. They provide ideas for customization and packaging. Therefore, blockchain resources allow electronic miners to collaborate and work closely together.

On the one hand, both blockchain cryptographic operations and non-crypto operations can access a set of computing resources such as the Mobile Edge Cloud (MEC). To improve the system's energy efficiency, Luo et al. [8] adjust the computing resources and block batch size, considering the trust characteristics of the SDN controller and the resource requirements of non-encryption operations. To realize a genuinely decentralized blockchain technique, we propose a partially observable Markov decision process (POMDP) and a new deep reinforcement learning (DRL) method to explain the problem and solve it. The simulation results compared three blockchain protocols and proved the effectiveness of each protocol scheme. Oktian et al. [9] propose a scalable, hierarchical blockchain architecture for the internet of things, consisting of a base engine and three auxiliary engines: payroll, computing, and storage stations. All of this makes it easy to run a workflow from IoT applications in documentation and proxies. They implemented a custom base engine and suggested the idea of using demand configuration and prioritization to improve its performance effectively. They also provided preliminary evidence for a conceptual implementation to evaluate engine-to-engine interoperability and implement concurrency in a pilot application for IoT car rental. Their evaluation results show that their proposal is feasible and works well in the local online environment.

To date, most IoT operating systems are cloud-centric and use an integrated platform provided by the cloud. However, cloud-based infrastructure primarily implements static sensor and data flow systems that do not support IoT components' direct configuration and management. To solve this problem, the virtualization of IoT components at the edge of the IoT network, Samaniego et al. [10], introduce an authentication-based blockchain protocol for providing virtualized resources directly to end devices. The proposed architecture focuses on deploying configuration tasks at the edge of an IoT network, using virtual resources and blockchain protocols as management tools. Their work also provides an evaluation of the implementation of the two permission-based blockchain protocol methods. The idea of connecting material things and networks to create new and productive interactions is a critical component of the concept of smart spaces. One of the significant challenges of these new intelligent spaces is controlling access to data, services, and things. Blockchain technology has become the most promising solution for distributed access management. Function-based access control allows users to access data/services/things by sending texts between decentralized account books. Of course, managing the access test transport path is a big deal. In IoT, smart contracts are the main component of access control in most blockchain network proposals [11]. One of the biggest problems with using smart contracts as verification of access codes that can be transferred from one account to another is that smart contracts and chain codes must be immutable by design since they represent a bond between the two parties.

IoT systems are made more attractive by enabling ubiquitous connectivity. More and more trends today are focusing on suspicious deals on smartphones. Another study by Samaniego et al. [12] looks at suspicious smart business transactions and studies blockchain technology's characteristics to manage them. Besides, their study reveals new concepts of blockchain operating systems and interactive contracts that will help explore how to navigate different markets in an unfamiliar environment through provenance.

3 Proposed Architecture

We have proposed a secure transaction management using blockchain as a service software for IoT. We have used Hyperledger as a private blockchain for simulation purposes. The transaction of Hyperledger are recorded securely and are publicly available. All transactions are digitally signed before transmission over the network [13]. So, the transaction is something more secure, which represents the authenticity and integrity of Hyperledger.

The following can be considered as properties of blockchain:

- Blockchain is a decentralized public decentralized, secure database between IoT nodes.
- Each IoT node has a function to verify a block.
- Some IoT nodes can be considered controllers of the blockchain.
- Peer-to-peer topology is being used in the blockchain.
- Hyper Ledger stores all recorded transactions. The blockchain will be updated after recording.

Figure 2 explains the client interaction model with blockchain and IoT platforms. IoT consist of different devices blockchain acts as the middleware between those devices and users of different subnetworks. It helps to manage the transaction between nodes, devices, and users securely.

Figure 3 shows the proposed architecture. It has two main parts one is IoT layers, and the other is blockchain structure. IoT layers are composed of collaboration, application, data abstraction, data accumulation, edge computing, connectivity, and physical devices. They send several types of information through sensors to the

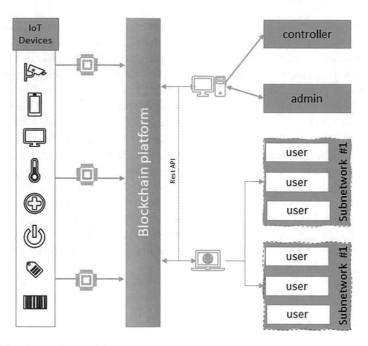

Fig. 2 Client interaction model

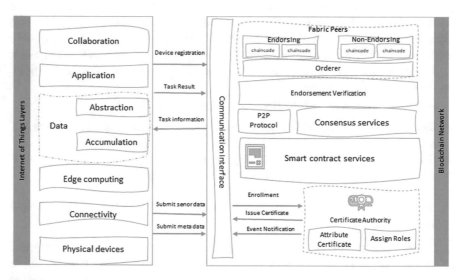

Fig. 3 Proposed architecture

blockchain. Blockchain interacts with these devices using the communication interface. Participants can have one or more peer nodes on the network. This structure defines several types of peer nodes on the model.

The first one is committing peer. As a key-value store, each peer maintains a current snapshot of the current state of the ledger. Such peers cannot call chaincode functions. The second one is endorser peer. Chaincode is installed on the Endorser peer. When they receive a transaction offer, they simulate transaction execution in an isolated container. Based on the simulation, such peers prepare transaction proposals and send them to the orderer's peers. Approving the existence of peers prevents all peers from executing transactions sequentially [14]. The third peer is called an orderer peer. The orderer receives the approved transaction and puts it in the block. After grouping the transactions, the orderer guarantees the agreement by propagating such blocks to the submitted peers, and the peers are verified and sent to the shared ledger. The orderer's peers record valid and invalid transactions, while other peers only include legitimate transactions [15].

4 The Transaction Flow

Transactions are transmitted in a peer-to-peer topology within the proposed framework. There are private IoT nodes in the network called miners and typically used to verify transactions. When the transaction is confirmed, it is converted into a block, added to the existing blockchain, and transmitted to the system. Miners play an essential role in coordinating newly created blocks on the blockchain. Follow the execution order confirmation form as follows.

4.1 Transaction Proposal

The blockchain client representing the organization writes a transaction proposal and sends it to colleagues identified in the referral policy [16]. The project includes the bidder ID, transaction burden and negligence, and transaction ID.

4.2 Endorsement

Recommendations include transaction simulation. Approvers create write and read groups that contain keys and modified values. The approval also verifies the legality of the execution of the transaction. Approval is sent after the proposal and includes the write package, read package, transaction ID, approver ID, and approver's signature [17]. When the customer has collected enough approvals (which should have the same result), a transaction is created and sent to the ordering service. The approval phase eliminates the final uncertainty.

4.3 Order

In this stage, the order is executed after approval. The requested service verifies that the blockchain client sending the transaction proposal has the rights (broadcast and receive rights) on the specified channel. The sorted block contains the approved parameters for each channel. With this request, the network can reach a consensus. The customer sends the results of the transaction to all colleagues.

4.4 Validate

First, each party confirms the transaction received by making sure the transaction matches its recommended strategy. Then it scans all the transactions in the block to read and write collisions in sequence. For each transaction, the read key view is compared to the current default ledger view. Make sure the values are the same. If they do not match, the other party rejects the transaction. Finally, update the ledger [18]. The ledger is based on the created block—additional ledger validation test results, including invalid transactions.

5 Performance Evaluation

This section presents the results of the proposed system evaluation.

Figure 4 shows the command-line interface for the starting the Hyperledger Fabric. After installation, we can use the Hyperledger composer for designing the underlying architecture of private blockchain. Table 1 shows the modeling environment used in the beta phase. Hyperledger Fabric Version 1.4.1 is used to define the operating system. The operating system used for modeling is Ubuntu Linux 18.04.1 LTS.

The Fig. 5 shows the transactions list of blockchain transactions. The user can view the immutable date and time of any transaction. It also shows the entry type and relevant participant. It allows the user to view the extended information of any record as well.

Figure 6 shows the bar graphs of these three groups. The network diagram shows that as the number of users increases, the delay changes. However, this does not affect performance. The average values for 300, 400, and 500 user groups are 148, 215, and 342 ms. And at least 73, 79, and 124. The maximum quality difference for a group of 500 users with a length of 561 ms. The Latency for the get request

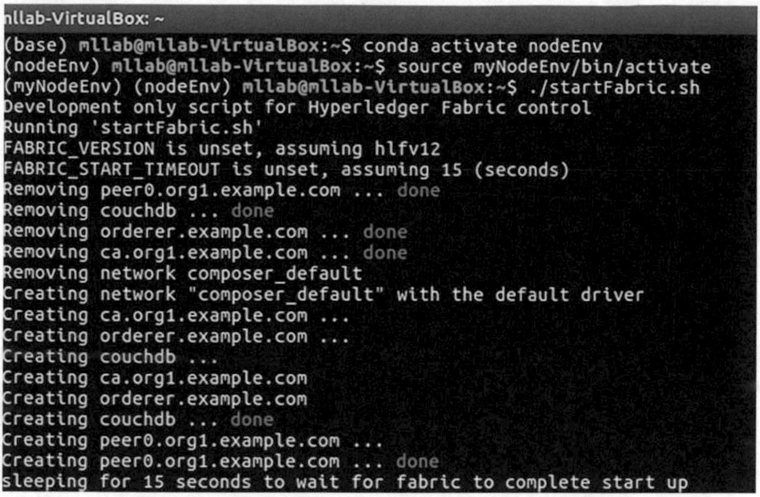

Fig. 4 Installation of Hyperledger Fabric

Table 1 Simulation setup

System component	Description
Operating system	Ubuntu Linux 18.04.1 LTS,
CPU	Intel ®Core ™ i5–8500 CPU at 3.00 GHz
Hyperledger Fabric	v1.4.1
Docker Engine	Version 18.06.1-ce
CLI Tool	composer-cli,
Docker-Composer	Version 1.13.0
Primary memory	16 GB RAM
Language	JavaScript
IDE (Platform)	Hyperledger composer-playground

transaction query evaluation results shows that the system will Indicates an increase in latency as the number of users increases [19].

6 Conclusion

This article uses a blockchain-as-a-service software package to describe and manage suspicious transactions in IoT systems. This research builds specific software components for blockchain functions to be implemented in IoT networks. This research is conducted using Hyperledger Fabric as a blockchain service to test software-defined blockchain components in the blockchain. We have used the transaction latency evaluation model to test the performance of the proposed system. We have confirmed that

Date, Time	Entry Type	Participant	
2020-06-24, 19:29:31	AddAsset	admin (NetworkAdmin)	view record
2020-06-24, 19:29:25	AddAsset	admin (NetworkAdmin)	view record
2020-06-24, 19:29:16	AddParticipant	admin (NetworkAdmin)	view record
2020-06-24, 19:29:12	AddParticipant	admin (NetworkAdmin)	view record
2020-06-24, 19:29:08	AddParticipant	admin (NetworkAdmin)	view record
2020-06-24, 19:29:04	AddParticipant	admin (NetworkAdmin)	view record
2020-06-24, 19:28:58	AddParticipant	admin (NetworkAdmin)	view record
2020-06-24, 19:28:49	AddParticipant	admin (NetworkAdmin)	view record
2020-06-24, 19:25:01	ActivateCurrentIdentity	none	view record
2020-06-24, 19:24:44	StartBusinessNetwork	none	view record

Fig. 5 List of transactions

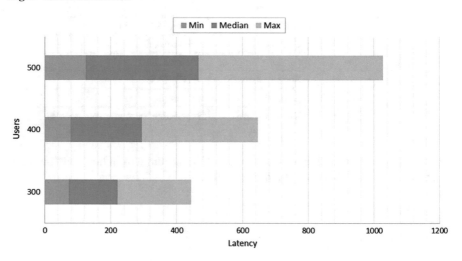

Fig. 6 Latency for the get request transaction query

using blockchain as a software service system can provide excellent performance and security. In the future, we aim to extend this work to do a practical implementation of the proposed method. We will also work on the specific use cases of the proposed transaction management system.

Acknowledgements This work was supported by Korea Institute for Advancement of Technology (KIAT) grant funded by the Korea Government (MOTIE) (N0002327, The Establishment Project of Industry-University Fusion District).

References

1. Novo O (2018) Blockchain meets IoT: an architecture for scalable access management in IoT. IEEE Internet Things J 5(2):1184–1195
2. Jiang Y, Wang C, Wang Y, Gao L (2019) A cross-chain solution to integrating multiple blockchains for IoT data management. Sensors 19(9):2042
3. Nyamtiga BW, Sicato JCS, Rathore S, Sung Y, Park JH (2019) Blockchain-based secure storage management with edge computing for IoT. Electronics 8(8):828
4. Jia X, Feng Q, Fan T, Lei Q (2012) RFID technology and its applications in Internet of Things (IoT). In: 2012 2nd international conference on consumer electronics, communications and networks (CECNet). IEEE, pp 1282–1285
5. Biswas S, Sharif K, Li F, Nour B, Wang Y (2018) A scalable blockchain framework for secure transactions in IoT. IEEE Internet Things J 6(3):4650–4659
6. Samaniego M, Jamsrandorj U, Deters R (2016) Blockchain as a service for IoT. In: 2016 IEEE international conference on internet of things (iThings) and IEEE green computing and communications (GreenCom) and IEEE cyber, physical and social computing (CPSCom) and IEEE smart data (SmartData). IEEE, pp 433–436
7. Samaniego M, Deters R (2019) Pushing software-defined blockchain components onto Edge Hosts. arXiv preprint arXiv:1909.09936
8. Luo J, Chen Q, Yu FR, Tang L (2020) Blockchain-Enabled software-defined industrial internet of things with deep reinforcement learning. IEEE Internet Things J
9. Oktian YE, Lee SG, Lee HJ (2020) Hierarchical multi-blockchain architecture for scalable internet of things environment. Electronics 9(6):1050
10. Samaniego M, Deters R (2018) Virtual resources & blockchain for configuration management in IoT. J Ubiquitous Syst Pervasive Networks 9(2):1–13
11. Samaniego M, Deters R (2018) Detecting suspicious transactions in iot blockchains for smart living spaces. In: International conference on machine learning for networking. Springer, Cham, pp 364–377
12. Samaniego M, Espana C, Deters R (2019) Suspicious transactions in smart spaces. arXiv preprint arXiv:1909.10644
13. Heilman E, Baldimtsi F, Goldberg S (2016) Blindly signed contracts: Anonymous on-blockchain and off-blockchain bitcoin transactions. In International conference on financial cryptography and data security. Springer, Berlin, pp 43–60
14. Sukhwani H, Wang N, Trivedi KS, Rindos A (2018) Performance modeling of hyperledger fabric (permissioned blockchain network). In: 2018 IEEE 17th international symposium on network computing and applications (NCA). IEEE, pp 1–8
15. Choudhury O, Sarker H, Rudolph N, Foreman M, Fay N, Dhuliawala M, Das AK (2018) Enforcing human subject regulations using blockchain and smart contracts. Blockchain Healthc Today 1:1–14
16. Dittmann G, Jelitto J (2019) A blockchain proxy for lightweight IoT devices. In: 2019 Crypto valley conference on blockchain technology (CVCBT). IEEE, pp 82–85

17. Khan PW, Byun YC, Park N (2020) A data verification system for CCTV surveillance cameras using blockchain technology in smart cities. Electronics 9(3):484
18. Khan PW, Byun YC (2020) Smart contract centric inference engine for intelligent electric vehicle transportation system. Sensors 20(15):4252
19. Hang L, Kim DH (2019) Design and implementation of an integrated iot blockchain platform for sensing data integrity. Sensors 19(10):2228

Plant Growth Measurement System Using Image Processing

Meonghun Lee, Haeng Kon Kim, and Hyun Yoe

Abstract Recently, improving the quality of life of farmers through the application of ICT in agriculture has attracted active attention. Smart farming refers to a system that cultivates or manages crops by integrating various sensors and cameras with computer and communication technologies. In this study, we developed a plant growth measurement system and designed a series of devices to increase system performance. Moreover, to improve the performance analysis based on environmental information and images, we used not only the faster R-CNN network to detect the location and size of pests but also a powerful tool to estimate the likelihood of identified pests. Furthermore, through this study, we have successfully advanced the existing artificial intelligence technology, thereby laying the foundation for AI in agricultural automation.

Keywords IoT · Application · Smartfarm · Image processing · Plant growth

M. Lee
Department of Agricultural Engineering, National Institute of Agricultural Sciences, Wanju-gun, Jeollabuk-do 55365, Republic of Korea
e-mail: leemh5544@gmail.com

H. K. Kim
School of Information Technology, Catholic University of Daegu, Gyeongsan, Gyeongbuk-do 38430, Republic of Korea
e-mail: hangkon@cu.ac.kr

H. Yoe (✉)
Department of Information and Communication Engineering, Sunchon National University, Suncheon, Jeollanam-do 57922,, Republic of Korea
e-mail: yhyun@sunchon.ac.kr

1 Introduction

Recently, improving the quality of life of farmers through the application of ICT in agriculture has attracted active attention. To manage crops, a farmer usually walks to the field to detect the growth of crops and trees with the naked eye. He/she then returns to an office to record the observation results, such as plant growth status, on separate sheets.

Drawbacks of this conventional method are highlighted. First, it is difficult for farmers to share information on crop management, and second, farmers can only rely on their experiences to handle the records of a particular crop and conditions of agricultural growth Additionally, the existing method is not suitable for overseeing a large area and a wide range of crops and trees at the same time. To address these issues, many studies have been performed on the application of ICT in agriculture to build a platform to enable the spread of smart farms [1–3].

Smart farming refers to a system that cultivates or manages crops by integrating sensors and cameras with computers and communication technologies. Smart farming researchers have been advancing different types of technologies, mainly sensor control, crop monitoring, disease prevention, and predicting harvest timing through database and detection. Furthermore, in the case of a crop management system implemented on a large-scale farm or a greenhouse, it is important to locate the spot at which the soil moisture content, temperature, and humidity are measured. However, since existing studies are unable to identify such locations, effective management of the system can only be achieved with the installation of the sensors at a predetermined place as well as prior registration of the site information. To solve this problem, we aim to build a monitoring system that can be controlled via a wired or wireless network using image processing solutions.

2 Background

2.1 Tomato Pest Management

The convergence and integration of ICT is an emerging alternative that can overcome the limitations of small farms and revive the agriculture industry in South Korea [4, 5]. Korean agricultural products are produced in farms that are often old and backward, which has led to a decline in the overall quantity and quality of the products. Thus, we believe that it is necessary to first modernize and systemize these facilities in order to increase productivity. Although horticultural management must be conducted in consideration of various environmental factors including temperature and insolation, this method is, however, individually controlled in South Korea, causing a low level of productivity in the agriculture industry.

Nevertheless, some advanced farms have been able to produce quality yields by collecting and diagnosing information on the growing conditions of crops, such as temperature, humidity, insolation, and carbon dioxide gas, through complex environmental control systems. Moreover, these farms have recently adopted information and communication technologies that enable the cultivation of crops as well as the creation of added value in logistics.

ICT in agriculture, which has attracted considerable attention recently, can create added value in production, distribution, and sales through the convergence of technologies of related industries and information gathering. With ICT, farmers can (1) quickly respond to environmental changes via interactive communication, (2) commercialize original ideas to increase their incomes, and (3) establish a production record system to achieve the convergence of different technologies in areas of agricultural production and management. A new era is approaching, in which farmers can conveniently and safely manage the greenhouse environment and produce their crops in smart farms.

ICT and its convergence in horticulture presents an opportunity for agriculture to move away from its production-oriented approach and converge with other industries. By creating added value and expanding the base of agriculture, this will serve as a stepping stone for South Korea to not only solve the existing issues in agriculture but also become an agricultural powerhouse.

2.2 Pest Prevention

Crop pests and diseases negatively affect the production of high-quality crops. However, if farmers spray pesticides to prevent such pests and diseases, further problems could arise as people become overexposed to harmful chemicals or pesticide waste, and environmental pollution can occur with excessive use. Moreover, because of a lack of information on new types of pests and diseases caused by abnormal climate phenomena, it is not easy for farmers, especially those migrating from the city to the countryside, to obtain accurate information even if they search for relevant texts and data. This has resulted in low productivity and yields of high-quality crops. To address this issue, we have introduced a system that can automatically control the amount of pesticides sprayed according to recommendations. Furthermore, to immediately provide information on plant pests and diseases, we have studied the pest information search system that allows users to search relevant images via their smartphones.

2.3 Environmental Control Through Database

The tomato is part of the nightshade family, which can be grown all year round in a greenhouse even though its main season is summer. To produce tomatoes continuously throughout the year, controlling the temperature and day length is essential, and to produce high-quality cut flowers, it is necessary to create an ideal environment with the optimal level of CO_2 concentration, humidity, nutrient solution, and irrigation. However, not only is it difficult to maintain such an environment but it also carries a high economic cost. Therefore, if the product value and ideal environment can be maintained until the shipping season, we believe this could greatly assist farmers to increase their incomes. Many researchers have studied tomatoes regarding their cultivation environment, including temperature, day length, and CO_2 concentration. Nonetheless, each of the many different types of tomato reacts to temperature and day length differently. In this study, we have established a plant growth measurement system to secure basic data for automation control and to create growth modeling.

3 Plant Growth Measurement System

3.1 Control Device

Master board. The master board is a board that receives sensor input from the driver board and sends control signals. The board is programmed with key input, sensor input values, and display processing, along with integrated control algorithms. To exchange information through the Wide Area Network (WAN), the master board communicates with the embedded board through the RS-232C method [6]. The control board in Fig. 1 is manufactured using Atmel's AT89C52 controller. This board uses ADC0809 for Analog-to-Digital Conversion (ADC). The features of the AT89C52 controller sensors are illustrated in Table 1 [7].

Although the controller shown in Fig. 1 is sufficient for information sensing, it lacks functions like CAN communication, AD conversion, and modification and supplementation for information exchange between different controllers. Hence, we have made a board, shown in Fig. 2, using Atmel's T89C51CC01 [8]. However, we have also found several bugs in the board due to an error in the circuit design, so we have rebuilt the board, as shown in Fig. 3. The board in Fig. 3 uses Liquid Crystal Display (LCD) as a display device and has CAN and RS-232C output ports. Moreover, since the board has an In-System Programming (ISP) function, it is easy to modify the program, which shortens its development time. To make the board

Fig. 1 Control board using AT89C52

Table 1 Features of the AT89C52 sensors

Sensors	Features
AT89C52	• Built-in 8 kB flash memory, 128 bytes of internal RAM • 32 programmable I/O lines (4 ports, each of which can control bitwise) • Built-in programmable serial port
ADC0809	• 8-channel input • Analog signal input range: 0–5 V • Resolution: 8-bit
LM35D	• Calibrated directly in Celsius • Linear +10.0 mV/°C scale factor • 0.5 °C Ensured Accuracy (at +25 °C) Low-Impedance Output, 0.1 Ω for 1 mA load

Fig. 2 Control board I using T89C51CC01

Fig. 3 Control board II using T89C51CC01

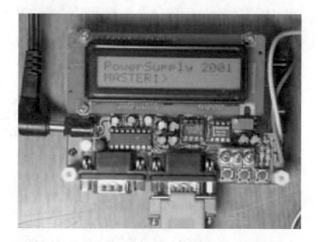

more generic, we have redesigned the circuit of Fig. 3, and finally created the last version of the master board shown in Fig. 4.

Fig. 4 Master board

We used OrCAD 9.2 for the circuit design, Keil uVision2 for controller programming, and FLIP 1.8.2 provided by Atmel for the ISP program [9].

Driver Board. The driver board consists of power, key input, display, sensor input, and control output, sending to or receiving information from the master board. The board has 16 channels for sensor input and another 16 channels as control signal output. The power supply section produces various power sources such as +5 V DC, ± 12 V DC, and +24 V DC to match the input power of each sensor. Although many existing devices use an LCD as a display device, in this study, we have used a 4 × 20 vacuum fluorescent display (VFD) as a display device. The reason is that when light reflects on the LCD surface, it is hard to discern what is displayed, and when the temperature is high, glares and halo occur. Conversely, VFD is resistant to temperature with excellent readability even when light is directly reflected on its surface.

Embedded Board. As for the embedded board, we have used Hyper104A produced by HyBus. Hyper104A (basic specification) is a high-performance development board created based on the Intel StrongARM SA1110 CPU [10]. Additionally, the operating system is made of an embedded Linux kernel, which makes fast and stable operation possible.

3.2   Communication Method

In this study, we used an RS-232C between the master board and embedded board while the CAN communication method was used between the master board and other master boards. We also used socket communications for the World Wide Web (WWW) communication.

RS-232C is an old standard used for physical connections and protocols for relatively slow serial data exchange between computers and related devices. The current version is "C." This standard is originally defined by the Electronic Industries Association (EIA), a group of industry companies for teletype devices [11].

CAN is a vehicle network system originally developed to provide digital serial communication between various instrumentation and control equipment in a vehicle. It was established as an international standard by the International Organization for Standardization (ISO) in 1993. CAN is a highly flexible network that supports master/slave, multiple masters, and peer to peer communications. It can withstand harsh environments of factories, high temperatures inside greenhouses, shocks, vibrations, and environments with significant noise. Thanks to these advantages, CAN has been widely used as a communication network for data exchange between control and automation-related equipment in various industrial facilities of factory automation and distributed control systems. Typical communication networks that use CAN include DeviceNet, SDS, CAN Kingdom, and CANopen/CAL. They all use CAN as their data link layer, but with different application layer protocols.

Sockets are a communication method between a client program and a server program on a network. Sockets are defined as "end of connection." They are created and used upon a function call or a programming request called "socket application program interfaces (APIs)." The most common socket API is the Berkeley Unix C language interface. Sockets are also used for communication between processes within the same computer.

3.3   IoT Sensors

Environmental factors affecting tomato growth include temperature, soil moisture, and CO_2. To obtain information regarding them, a sensor suitable for each element is required. In this experiment, we have used a temperature sensor, soil water tension sensor, and CO_2 sensor to collect environmental information in the greenhouse. Since the environment inside the greenhouse is different from and superior to the outer one, we chose each sensor with a proven record of accuracy, stability, and reliability.

As for the temperature sensor, we have used an AD592 with 5 V DC input power. This sensor receives a current output by connecting a precision resistor at the end in parallel with the ground.

As for the soil water tension sensor, we have used an SKM850 (SDEC) with 12 V DC input power. This sensor has a differential output voltage. Therefore, the differential voltage is amplified to a certain range by obtaining the difference between the two signals through the differential amplifier. If the differential amplifier is not used, it needs two channels to receive the sensor input. However, if the differential amplifier is used, the input can be received through one channel. The differential amplifier we used is the Burr Brown's INA103.

As for the CO$_2$ sensor, we have used a GMW20 (VAISALA) with 24 V DC input power. This sensor receives a current output by connecting a precision resistor at the end in parallel with the ground. The measurement voltage range is 0–3 V.

3.4   DB Server

DB is a collection of data that is structured in such a way that its contents can be easily accessed, processed, and updated. The most widely used DB is a relational DB, a table-type DB that can be reconstructed and accessed again in various ways. Distributed DB refers to a DB that is distributed or overlapped at several different points on the network, whereas object-oriented DB refers to a DB in which data defined as an object class and a subclass coincide with each other. In this study, we have used a MySQL Version 5.0.77, which is a relational DB. Using C API and JDBC, MySQL can easily develop programs through C programming or Java [12].

The entity-relationship diagram created in this way is converted into a schema diagram according to the relational schema conversion rule. Shown in Fig. 5, the composition of the completed schema diagram is as follows.

- Board: the table that manages information related to the bulletin board
- Board_kind: the table that manages the grade according to the type of bulletin board

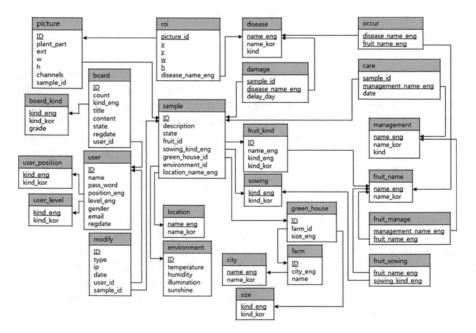

Fig. 5 Schema diagram

- Care: the table that connects the sample table and management table; it enables you to see which control and nutritional agents were used for the sample
- City: the table that manages the location of the farm and stores the information of local governments
- Damage: the table that connects the disease table and sample table; it also enables you to see which diseases and pests have occurred in the sample
- Disease: the table that manages illness information divided into pests and diseases
- Environment: the table that manages greenhouse information such as temperature, humidity, sunlight, and illuminance
- Farm: the table that manages the location and name of the farm
- Fruit_kind: the table that manages crops according to their types
- Fruit_manage: the table that connects the fruit_name table and management table. One can find the types of pest control and nutrients corresponding to crops
- Fruit_name: the table that manages names of crops. Currently, there is only one type of tomato, but you can add other crops as needed
- Fruit_sowing: the table that connects the fruit_name table and sowing table where the seeds of particular types of crops can be found
- Green_house: the table that manages the size of the greenhouse and the farm the greenhouse belongs to
- Location: the table that manages the location of samples caused by different types of outbreaks, such as the outside, center, or vents of the greenhouse
- Management: the table that manages types of pests
- Modify: the table that manages data modification information and manages information of users who modify data such as ID and IP address.
- Occur: the table that connects the fruit_name table and disease table, where it is possible to see which diseases and pests occur in which crops
- Picture: the table that manages video information such as size, channel, and extension
- Roi: the table that manages information on the region of interest where the disease appears on images
- Sample: the most important table in the database system, which manages comprehensive data describing the disease and pest images, shooting environment, shooting location, seeding type, and other peculiarities
- Size: the table that manages the size of the greenhouse
- Sowing: the table that manages the types of sowing
- User_position: the table that manages the positions of the types of user, such as expert, farmer, professor, student, and the general public
- User_level: the table that manages the types of user levels such as administrator, beginner, intermediate, and advanced
- User: the table for managing user accounts such as name, password, title, and level.

4 Plant Growth Measurement System Using Image Processing

After analyzing the disease and pest recognition and detection based on deep learning, we have finally selected the region-based convolutional neural network (CNN) as the architecture. A network of this kind can detect the location and size of pests in an image, and uses a powerful method to calculate the probability of identified pests. We have chosen the faster R-CNN as our neural network, which is an extended architecture with high speed and has all the advantages of R-CNN, SPPNet, and fast R-CNN [13, 14].

Since the size of the input image is fixed on the existing CNN, users transform or crop the images, ignoring their original ratios when they want to use them. In this case, the information included in the image may be damaged, but as for the Faster R-CNN, the size of the image is irrelevant, which makes it immune to the problems caused by camera type and image size. The region-based CNN has improved the performance of the object recognition by extracting the region where the object is likely to exist through the region proposal process. However, this method requires a large amount of computation, which is why faster R-CNN through fast R-CNN was developed to improve the speed. Furthermore, environmental data are annotated and combined with environmental information at the time of image data collection.

In the region proposal stage, faster R-CNN uses a region proposal network (RPN) instead of selective search (SS). The network creates a feature map through CNN's feature extraction capability. Subsequently, the RPN generates a feature vector including the coordinates and probability of the bounding box through a sliding window process. ROI pooling is then performed from the sampled feature codes. The extraction is finally completed upon identifying the probability of each class during the classification process. Figure 6 demonstrates the process of detecting a pest through the process of extracting and classifying bounding boxes in faster R-CNN. Table 2 illustrates the number of bounding boxes and average classification accuracy for each pest type learned by Faster RCNN.

5 Conclusions

With the development of deep learning neural network technology that can recognize objects at the human level by learning images in the same way as humans, it is possible to revolutionize the automation/artificial intelligence of agriculture. This study attempted to develop a learning engine of the deep learning neural network and to advance a plant growth measurement system based on image processing through deep learning of big data of tomato disease and pest images. This study has also laid the foundation for the artificial intelligence of agricultural automation. Although the fast R-CNN used in this study improved the performance remarkably, in terms of speed and accuracy, by integrating R-CNN's complex training/test pipeline, a further

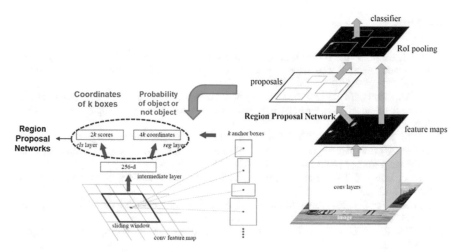

Fig. 6 Faster R-CNN operation process

Table 2 Number of bounding boxes generated by the network and the average classification precision of the test set	Type of pests	Number of bounding boxes	Percentage [%]	Average precision
	Leaf Mold	11,921	27	0.9061
	Gray Mold	2765	6	0.7969
	Bacterial Canker	2645	6	0.8560
	Phytophthora Root Rot	2573	6	0.8762
	Leaf Miner	2943	7	0.8046
	Cold Damage	477	1	0.7824
	Powdery Mildew	332	1	0.6556
	Whitefly	402	1	0.8301
	Nutrient Burn	421	1	0.8971
	Background	18,896	44	0.9005
	Mean A	43,397	100	0.8307

improvement can still be made in terms of the processing speed to get one step closer to the real-time object detector. In the future, we plan to dramatically reduce the inaccuracy and time delay to increase the precision of image-based pest recognition by using various network models.

Acknowledgements This research was supported by the Ministry of Science and ICT (MSIT), Korea, un-der the Grand Information Technology Research Center support program (IITP-2020-0-01489) supervised by the Institute for Information and Communications Technology Planning & Evaluation (IITP).

References

1. Lee M, Hwang J, Yoe H (2013) In Agricultural production system based on iot. In: 2013 IEEE 16th international conference on computational science and engineering, 3–5 Dec 2013, pp 833–837
2. Lee M, Kim H, Yoe H (2017) In intelligent environment management system for controlled horticulture. In: 2017 4th NAFOSTED conference on information and computer science, 24–25 Nov 2017, pp 116–119
3. Lee M, Kim H, Yoe H (2019) Icbm-based smart farm environment management system. In: Lee R (ed) Software engineering, artificial intelligence, networking and parallel/distributed computing. Springer International Publishing, Cham, pp 42–56
4. Kim G, Huh J (2015) Trends and prospects of smart farm technology. Electron Telecommun Trends 30:1–10
5. Kim J-T, Han J-S (2017) Agricultural management innovation through the adoption of internet of things: Case of smart farm. J Digital Convergence 15:65–75
6. Lascano RJ (2000) A general system to measure and calculate daily crop water use. Agron J 92:821–832
7. Song Y, Feng Y, Ma J, Zhang X (2011) Design of led display control system based on at89c52 single chip microcomputer. JCP 6:718–724
8. Krastev G (2004) Microcomputer protocol implementation at local interconnect network. Comput Appl 24:114–116
9. Rashid MH, Rashid MH (2004) Introduction to pspice using orcad for circuits and electronics. Pearson/Prentice Hall
10. Corporations I (2000) Intel strongarm sa-1110 microprocessor develper's manual. June: 2000
11. Cirincione J, Bacharach, S (2007) Data standards and service standards: Helping businesses in real estate, mortgage, appraisal, and related industries function more efficiently. J Real Estate Lit 127–137
12. DuBois P (2008) Mysql. Pearson Education
13. Girshick R (2005) In Fast r-cnn. In: Proceedings of the IEEE international conference on computer vision, pp 1440–1448
14. Purkait P, Zhao C, Zach C (2017) Spp-net: Deep absolute pose regression with synthetic views. arXiv preprint arXiv:1712.03452

Smart Cattle Shed Monitoring System in LoRa Network

Meonghun Lee, Haeng Kon Kim, and Hyun Yoe

Abstract In this paper, we describe the design of "smart cattle shed" monitoring system based on LoRa, a low-power, long-range wireless communication technology, to support the range of communications and safety measurements required when operating large-scale smart cattle shed. The proposed system wirelessly collects real-time stable information from sensors installed in the cattle shed, and the collected data are analyzed by the integrated management system, delivered to the user, and controlled by the application. We also used network encryption to enhance the security of our network, and a smart cattle shed monitoring algorithm was developed to control the optimal cattle shed environment. To secure the data transmission reliability of the sensor, a LoRa-based wireless communication was constructed and the communication performance was measured. The proposed system reduces the cost of smart cattle shed management by using a LoRa wireless network, and improved the safety of data transmission.

Keywords Smart cattle shed · LoRa · Wireless · IoT · Sensors

M. Lee
Department of Agricultural Engineering, National Institute of Agricultural Sciences, Wanju-gun, Jeollabuk-do 55365, Republic of Korea
e-mail: leemh5544@gmail.com

H. K. Kim
School of Information Technology, Catholic University of Daegu, Gyeongsan, Gyeongbuk-do 38430, Republic of Korea
e-mail: hangkon@cu.ac.kr

H. Yoe (✉)
Department of Information and Communication Engineering, Sunchon National University, Suncheon, Jeollanam-do 57922, Republic of Korea
e-mail: yhyun@sunchon.ac.kr

© The Author(s), under exclusive license to Springer Nature Switzerland AG 2021
H. Kim and R. Lee (eds.), *Software Engineering in IoT, Big Data, Cloud and Mobile Computing*, Studies in Computational Intelligence 930,
https://doi.org/10.1007/978-3-030-64773-5_12

1 Introduction

Given the rapid development of IT and sensor network technologies, studies examined the creation and development new services that incorporate these technologies to a variety of industrial sectors [1, 2].

The livestock industry, which was historically labor intensive, has progressed owing to the convergence of a diverse range of information technologies. With these advanced technologies, the industry is undergoing a transformation to enable stable and sustainable growth, and is evolving beyond its conventional pattern of high-risk and nature-dependent characteristics. However, these technologies are still under development, and most cattle sheds are failing to keep pace with new trends. Consequently, as abnormal weather conditions, such as heat waves, cold waves, and heavy rainfall are becoming the norm, the damage to climate-sensitive livestock has increased rapidly.

To reduce damage from the natural disasters and the incidence of fatal diseases to livestock, an optimal growth environment is necessary; to this end, a monitoring system is required that enables tracking of the cattle shed environment [3].

Most existing cattle shed facilities are conventional environments that lack up-to-date technologies: the cattle shed manager has to check everything visually and in person, which is labor-intensive and time-consuming. Furthermore, as it is not possible to achieve comprehensive monitoring of the livestock farming environment of the cattle shed, it is difficult to achieve the optimal growth environment for the livestock. Science and technology in South Korea are developing as fast as in developed countries, but the level of adoption of convergence technology in the livestock sector is still low, and there is a pressing need for the dissemination of such technologies. In addition, as there are few examples of introducing environmental management at the housing stage and of proper implementation, a test case that can be adopted as an environment management model is necessary.

Therefore, it is necessary to develop a model appropriate for the real-world situation of livestock farms in South Korea through a thorough analysis and exploration of the risk factors at the housing stage that may have a significant impact on the livestock housing environment and safety based on the livestock housing environment in Korea.

To overcome the current difficulties in cattle shed management, in this paper, we have proposed a wireless sensor network-based cattle shed monitoring system to address these problems. The proposed system is composed of various environmental sensors and wireless sensor network technology; the system collects environmental information on the cattle shed through an environmental sensor, transmits and analyzes the information wirelessly, and informs the manager of the current status of the cattle shed.

This system enables real-time monitoring of the cattle shed environment and proactive responses and actions depending on the situation; thus, it is expected that damage to livestock farms in emergency situations can be prevented. Additionally, it is possible to ease the management process for cattle shed managers, and move away from the manual methods used in the past, which involved frequent visits to the cattle shed and visual inspection of the maintenance and management of the cattle shed environment.

2 Related Research

2.1 Cattle Shed Monitoring Systems

For the existing environmental monitoring system, there has been a case study applied to the agricultural environment [4]. In this case, an environmental monitoring system was constructed using sensors for temperature, humidity, leaf temperature, and leaf wetness, and information on the greenhouse environment and crop conditions was collected through sensors, stored in a database, and the condition of the greenhouse was monitored through the internet. This study introduces an early-stage environmental monitoring system, and it is difficult to demonstrate whether the monitoring has been conducted properly because there are fewer environmental sensors. These features are still under evaluation, and we expect that the environmental information provided is somewhat insufficient for use in the actual cattle shed environment. Therefore, in this study, we have proposed a monitoring system that supports an effective system that enhances optimal growth environment by including the necessary sensors for the collection of cattle shed environment information.

2.2 IT Application in Livestock Industry

The United States, Canada, Japan, Australia, and the European Union have enacted regulations on quality and hygiene management of livestock products and applied them to livestock production. In recent years, development of the livestock industry and the creation of new business have been achieved through convergence projects promoting the use of various information technologies. In particular, the introduction of new technologies, such as livestock growth monitoring, livestock product distribution support, and disease monitoring systems, is highly effective, and a model that can implement integrated convergence services has been selected and intensively promoted. In these projects, state-of-the-art convergence technologies, such as RFID, USN, telematics, CCTV, and GIS/GPS, are actively utilized, and the companies that have won the projects are developing a service model that can integrate and link the entire processes of production, distribution, sales, and management in the livestock sector. Given the increasing number of livestock product processing plants, the potential to enhance competitiveness has been shown in practice.

2.3 IoT Environmental Sensors

The sensors required for cattle shed monitoring were set for temperature, humidity, CO_2, NH_3, and H_2S. Temperature sensors are employed depending on the intended purpose of use, such as temperature range detection and precision, temperature characteristics, mass productivity, and reliability. In the industrial sector, thermocouples, resistance thermometers, and metal thermometers are used. For household products, thermistors and temperature-sensitive ferrite thermometers are often used, and for special purposes, sensors using ultrasonic waves and optical fibers are also utilized. In essence, it is used to respond to changes in temperature and for automation of temperature management by detecting changes in temperature.

A humidity sensor is used to detect humidity, and makes use of various phenomena related to moisture in the air. Specifically, psychrometers, hair hygrometers, lithium chloride humidity sensors, and ceramic humidity sensors, are used depending on the field of application. Essentially, it is used to detect humidity in the air by converting it into an electrical characteristic value.

A CO_2 sensor is used to measure CO_2 concentration. The main types of CO_2 sensors are infrared gas sensors and chemical gas sensors. The measurement of carbon dioxide is important for monitoring indoor air quality and multiple industrial processes, and is used widely used in air conditioners as a means of monitoring air quality.

An NH_3 sensor is used to measure the NH_3 concentration. It is composed of an NH_3 ion-selective membrane or an NH_3 gas-permeable membrane and a composite hydrogen ion electrode, and is based on the principle of measuring the change in hydrogen ion concentration caused by NH_3 ions or NH_3 gas diffused through the membrane with the hydrogen ion electrode. As the sensor is sensitive to NH_3 and hardly responds to other volatile substances, it is used to measure NH_3 for cases such as industrial wastewater or NH_3 in cattle sheds.

H_2S sensors are used to measure toxic and hazardous gases. The gas concentration is measured continuously through the energy generated by the redox reaction, and the converted concentration is sent as a signal. This method is used to detect and measure complex odors generated in industrial complexes and odor emission areas (Tables 1 and 2).

3 Design of an Efficient Data Collection System

3.1 IOT Node for LoRa

The system for cattle shed environment data acquisition and positioning is designed based on LoRa communication [5, 6]. The cattle shed internal/external environment data acquisition and positioning system should include a LoRa communication module for data communication with a server, a Beacon signal receiver for

Table 1 Pigsty environmental sensors (internal sensors)

Name of device	Name of components	Photography	Description
Internal sensors	Temperature sensor		• Measurement of temperature inside the cattle shed • Temperature: -30 to $60\ ^\circ\text{C}$ (accuracy 0.2, based on Celsius) • Display unit: $\pm 1\%$ rdg, ± 1 digit • Consideration of durability and sensor life required
	Humidity sensor		• Measurement of moisture content in the air inside the cattle shed • Measurement range: 0–99% (accuracy 1%) • Display unit: $\pm 3\%$ rdg, ± 1 digit • Consideration of durability and sensor life required
	NH_3 sensor		• Measurement of NH_3 concentration inside the cattle shed • Measurement range: 0–3000 ppm • Consideration of durability and sensor life required
	CO_2 sensor		• Measurement of CO_2 concentration inside the cattle shed • Measurement range: 0–5000 ppm (Accuracy ± 30 ppm $\pm 5\%$) • Consideration of durability and sensor life required
	Blackout sensor		• 220–380 V • The blackout sensor and short circuit sensor are similar, but they are independent • Consideration of durability and sensor life required • The measurement method is different depending on the manufacturer

(continued)

Table 1 (continued)

Name of device	Name of components	Photography	Description
	Fire sensor		• Heat: $-20\ ^\circ\text{C}$ to $150\ ^\circ\text{C}$/Smoke: 0–100% • Measurement range: $\pm 0.5\ ^\circ\text{C}$/ $\pm 0.5\%$ • Consideration of durability and sensor life required • The measurement method is different depending on the manufacturer

indoor positioning, and a GPS signal receiver for outdoor positioning. In addition, the essential component of the system includes a three-axis acceleration sensor to detect danger, and an MCU to read and process data from each module.

A risk detection module and a positioning module are included in the system, and after compiling this data at the MCU, a LoRaUplink message is generated and the message is sent through the LoRa communication module. The system can be described as a battery-operated LoRa terminal. When the power is turned on, the MCU functions, and environmental data acquisition and positioning are performed at the same time. While the MCU is in operation, it receives data from the positioning module. When the positioning module operates initially, it starts to receive GPS signals, and when no data are received for a certain period of time, it determines that it is an indoor situation and starts receiving Beacon signals. Additionally, it attempts to receive a GPS signal at predetermined time points, and does not change the state if no data are received. When a GPS signal is received, it determines that it is an outdoor situation and continuously receives the GPS signal. In the risk detection module, inference is performed using pattern analysis.

A mathematical formula was constructed in which the square root of the result of addition of all signal data from each axis was determined to be the value of gravity, and when a free-falling signal was detected, it was determined to be a risk signal. In addition, risk data are used to assess the situation of risk.

The judgment of a risk situation is made based on two threshold values, and risk is determined when data are detected in the order of lower - > upper.

The MCU receives data from both the environmental data acquisition module and the positioning module, and then generates a LoRa Uplink message, which generates different messages according to the state.

When transmitting data from the MCU, the LoRa 904.3–905 MHz band is used, and in the case of transmission via channels, 8 channels, from 8 to 15, should be used.

As for the specification of the LoRa message, data communication proceeds according to v.1.0.2, and the message payload proceeds according to the basic specification of Semtech. The data transmitted from the MCU are configured to one of two conditions. For outdoor conditions, data are transmitted by collecting coordinate

Table 2 Pigsty environmental sensors (internal sensors and others)

Name of device	Name of components	Photograph	Description
External sensors (ambient atmosphere)	Temperature sensor		• Measurement of temperature outside the cattle shed • Temperature: −30 to 60 (accuracy 0.2, based on Celsius) • Display unit: ±1% rdg, ±1 digit • Consideration of durability and sensor life required
	Humidity sensor		• Measurement of moisture content in the air outside the cattle shed • Measurement range: 0–99% (Accuracy 1%) • Display unit: ±3% rdg, ±1 digit • Consideration of durability and sensor life required
	Wind sensor		• Measurement of wind direction and velocity outside the cattle shed • Wind velocity measurement range: 0–30 m/s • Wind velocity measurement time: 5 m/s • Wind direction measurement range: 0–359° • Minimum measurement time: 5SECm/s • Consideration of durability and sensor life required
Other devices	Lightening protection (Surge protective device)		• A device protecting electrical/electronic equipment or facility for livestock farming from overvoltage/overcurrent and lightning • Prevent random damage caused by lightning

(continued)

Table 2 (continued)

Name of device	Name of components	Photograph	Description
Integrated controllers	Controller		• A device that comprehensively controls the data measured by each sensor
	Data collector		• A device that collects data measured from internal/external sensors
	Communication relay		• A device that connects and manages communication between each sensor and controller, and between the controller and PC

data using a GPS signal; for indoor conditions, Beacon-related data, such as Beacon ID and RSSI can be transmitted using the Beacon signal.

The number of message transmission bytes for the Beacon signal and the GPS signal message transmission is specified as 11 bytes for each message. The second packet is omitted immediately after the STX packet becomes the message structure packet, and the data are classified and processed in the server using the second packet.

In case of the message transmission specification for the Beacon signal, information on the nearest Beacon can be acquired for transmission. As for the emergency packet, when detecting fall data, the transmission includes the data and the battery data from the current terminal.

In case of message transmission specification for GPS signals, the coordinates of the current position obtained using the GPS signal receiver are transmitted. Data are transmitted in units of degrees, minutes, seconds, and hundredths of a second, so that the current coordinates can be ascertained in detail.

3.2 Optimum Growth Environment for Livestock

The optimal environmental conditions for an actual pigsty are shown in Table 3. These conditions were set in reference to the environmental conditions with good productivity when raising pigs as presented by the National Institute of Animal Science.

Table 3 Optimum growth environment for livestock

Sensors	Optimal environment conditions
Temperature	15–20 °C
Humidity	60–65%
CO_2	Under 500 ppm
NH_3	Under 20 ppm
H_2S	Under 20 ppm

- **The temperature** needs to be lowered by ventilation or an air conditioner during the daytime to meet the optimal growth environmental conditions, and it is necessary to raise the temperature by heating owing to a sharp drop in temperature at night-time.
- **Humidity** remained constant on average; if necessary, it should be controlled using ventilation and water. When housing pigs, the correct humidity is 60–65%: if the humidity drops, the animals may develop respiratory diseases, so the humidity should be carefully controlled. CO_2 rises and falls with temperature. When heating to raise the temperature, the amount of CO_2 increases rapidly.
- **CO_2** is not a harmful gas that poses a direct risk to livestock, but if it is increased, the amount of oxygen decreases and livestock may die. It is also a useful indicator of the air quality inside the cattle shed.
- **NH_3** is a gas that may cause health problems in livestock. High concentrations of NH_3 lead to various symptoms, including headache, vomiting, and loss of appetite, and long-term inhalation causes damage to the cilia, which remove foreign substances from the airway.
- **H_2S** is likely to be generated from livestock manure. H_2S is a colorless, poisonous gas with an odor. If the concentration exceeds 700 ppm, it is toxic, and can cause an instant loss of consciousness and death after just one or two inhalations; thus, H_2S concentrations should be monitored and controlled.

4 System Evaluation

4.1 Transmission of Sensor Information

Sensor information is acquired at the request of the hub. The hub sends a sensor value request message from the sensor to the IoT node at a certain interval (e.g., every 1 min). The local ID of the desired sensor is used as a factor. The IoT node receives the request and sends the final acquired value and the local ID of the sensor in the sensor value response message. The final acquired value of the sensor is an analog value (or in the case of a digital sensor, is converted to an analog value for transmission according to the virtual linear rule) [7] (Fig. 1).

The hub receives the sensor value response message and publishes the sensor value to the cloud. The cloud stores the sensor values in a database. In the case of a

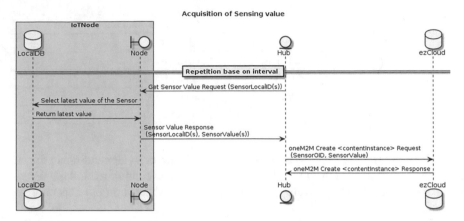

Fig. 1 Transmission of sensor information

digital sensor, the sensor value transmitted is converted back to a digital sensor value to be saved when stored in the database in the cloud.

4.2 Control System

A user can give a control command to the controller through the Cloud's IoT equipment control system. In addition, Cloud can send a control command to the controller through its own algorithm [5] (Fig. 2).

The ID for the control command and the control command are sent from the Cloud to the hub, and when the hub receives the control command, a control request message is sent with the ID to the control command, the local ID of the controller, and the factor to be controlled to the IoT node connected to the controller.

The IoT node starts controlling the applicable controller and sends a control response message along with the ID for the control command. Here, the control

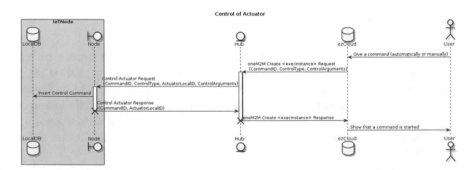

Fig. 2 Control system

response does not indicate that the control command has actually been started or completed, but that the control command has been successfully transmitted.

When the hub receives the control start message, it publishes an ID for the control command and a message indicating the start of control to the Cloud. The Cloud may or may not take appropriate action on the start of the control.

5 Conclusions

In a conventional cattle shed monitoring, to check the temperature and humidity in each area, a thermometer and a hygrometer are installed, respectively, and the values must be visually checked, in person, at the site. In addition, the accurate control of humidity is difficult and the process is time-consuming and labor-intensive, as frequent visits are required for the maintenance and management of the cattle shed environment. The system proposed in this study enables a comprehensive monitoring of the internal environment following installation of sensors for each area inside the cattle shed. Furthermore, monitoring is possible through PCs and smart devices without direct visits to the cattle shed, providing convenience to the cattle shed managers. Moreover, by applying an efficient data acquisition method based on LoRa, data quality and transmission/reception distance were improved compared with the conventional methods. To obtain precise information of the cattle shed environment, various environmental sensors and an efficient data acquisition method were used to implement an effective system, than the one that exists. Additionally, it was possible to check and examine the performance of the system through verification using a test-bed, and it is expected that more precise monitoring will be possible with more data in the future. The proposed system is expected to achieve eco-friendly agricultural production and scientific livestock farming with lower costs and greater efficiency by monitoring various environmental factors that influence the optimal growth environment in the cattle shed, such as temperature, humidity, and various gases.

Acknowledgements This research was supported by the Ministry of Science and ICT (MSIT), Korea, un-der the Grand Information Technology Research Center support program (IITP-2020-0-01489) supervised by the Institute for Information & Communications Technology Planning & Evaluation (IITP).

References

1. Bae J, Lee M, Shin C (2019) A data-based fault-detection model for wireless sensor networks. Sustainability 11:6171
2. Lee M, Hwang J, Yoe H (2013) In agricultural production system based on iot. In: 2013 IEEE 16th international conference on computational science and engineering. 3–5 Dec 2013, pp 833–837

3. Jung J, Lee M, Park J (2019) "Livestock disease forecasting and smart livestock farm integrated control system based on cloud computing", Korean institute of smart media. Korean Inst Smart Media 8(3):88–94. https://doi.org/10.30693/smj.2019.8.3.88
4. Lee M, Eom K, Kang H, Shin C, Yoe H (2008) In design and implementation of wireless sensor network for ubiquitous glass houses. In: Seventh IEEE/ACIS international conference on computer and information science (icis 2008), 14–16 May 2008, pp 397–400
5. Kim S, Lee M, Shin C (2018) IoT-based strawberry disease prediction system for smart farming. Sensors 18:4051
6. Ma R, Teo KH, Shinjo S, Yamanaka K, Asbeck PM (2017) A gan pa for 4 g lte-advanced and 5 g: Meeting the telecommunication needs of various vertical sectors including automobiles, robotics, health care, factory automation, agriculture, education, and more. IEEE Microwave Mag 18:77–85
7. Hwang JH, Lee MH, Ju HD, Lee HC, Kang HJ, Yoe H (2010) Implementation of swinery integrated management system in ubiquitous agricultural environments. J Korean Inst Commun Inf Sci 35(2B):252–262

The Triple Layered Business Model Canvas for Sustainability in Mobile Messenger Service

Hyun Young Kwak, Myung Hwa Kim, Sung Taek Lee,
and Gwang Yong Gim

Abstract Due to the spread of digital technology, new business models are emerging and the importance of sustainable growth for companies is being emphasized. Sustainability should consider both the environmental and social aspects, not just the economic aspect. It has been confirmed that digital technology has a positive effect not only on economic performance but also on environmental and social aspects. By applying it to the service, we were able to confirm that digital technological innovation can achieve corporate sustainability.

Keywords Sustainability · Triple layered business model canvas · Business model innovation · Messenger · Kakao

1 Introduction

In the contemporary society, new mobile services based on smart devices are appearing due to the development of digital technology and wireless communication technology. Moreover, due to the generalization of digital technology, function and services provided by the leading company became vulnerable to imitation and the importance of providing distinct customer value has increased. In addition, as

H. Y. Kwak · M. H. Kim
Departmentof IT Policy and Management, Soongsil University, Seoul, South Korea
e-mail: khykhan@naver.com

M. H. Kim
e-mail: beauhwa1@naver.com

S. T. Lee
Department of Computer Science, Yong In University, Yongin, South Korea
e-mail: totona22@yongin.ac.kr

G. Y. Gim (✉)
Department of Business Administration, Soongsil University, Seoul, South Korea
e-mail: gygim@ssu.ac.kr

social and environmental problems such as economic inequality, employment insta-
bility, environmental problems, and energy problems occur, problems related to the
external corporate environment are now perceived as a must-be-solved problem. If
not, they often threaten even the survival of corporation [1]. In order for a company
to create an innovative value for the customer and grow itself while providing the
value constantly, the company should consider not only the economic interest but also
the social and environmental responsibility when creating a new creative business
model.

Until the present day, studies on business model used Osterwalder's 'business
model canvas' methodology as the basis to suggest some variant models or analyzed
the success factors of a product or service by changing the components of the frame
[2, 3]. Based on this, these studies sought for new opportunities for innovation.
However, it focused solely on the analysis on the economic value of a company and
was not able to integrate today's complex business environment [1].

The purpose of this study is to define the concept of sustainability by applying
the 'triple layered business model canvas', which can analyze economic, environmental, and social aspects at the same time, to messenger service cases. I would like
to confirm that the services provided by companies can create economic value while
creating new business opportunities in both environmental and social aspects [1].
Finally, while confirming the implications and limitations of this analysis methodology, through a case study of kakaotalk, I would like to present the implications that
kakaotalk needs for continuous value creation in the future.

2 Sustainability

2.1 Sustainability and Triple Bottom Line Model

Recently, global corporations have been showing great interest in the concept of
sustainability. They have been talking about 'sustainable development' in the fields
of politics, business, economy and society as an important long-term plan for not only
the present but also the future. This shows how sustainably has become a leading
paradigm that leads the changes in the twenty-first century human society and a
central element in modern culture [1].

Sustainability generally means the ability to maintain a specific process or ecology.
Moreover, it refers to the new approach that seeks to pass on a long lasting and better
environment for the future generation [1]. Sustainability is also described as sustainable development (SD) by the United Nations and refers to a set of methods to protect
the environment, save poverty, and create economic growth without destroying
short-term natural resources.

In the early days, this concept of sustainability originated from an environmental
crisis. At the United Nations Conference on Human Ecology in Stockholm in 1972,

the attention of industrial development-oriented developing countries to the environmental protection because it was emphasized that the environmental destruction caused by industrial development is related to the problem of wealth concentration and poverty. In the report "Brundtland report: Our common future" published by the World Commission on Environment and Development (WCED) in April 1987, the concept of 'sustainability' was first formulated, and this report highlighted the importance of 'sustainable development' by defining sustainable development as 'an evolution that meets the present need within the capacity to meet the needs of future generations' [1, 4].

In 1992, the UN Conference on Sustainable Development in Rio de Janeiro discussed concrete measures for realizing economic, environmental and social sustainability, and as a result of the meeting, Agenda 21 was adopted and The United Nations Commission on Sustainable Development (UNCSD) was established. Since then, people started to understand sustainability as a concept that should be accompanied by three items: integration of economic growth, environmental protection, social development.

In these situation, John Elkington proposed a triple bottom line (TBL) as a research model of sustainability [5]. According to Elkington, a company can consider not only economic value but also environmental and social value through this model. Elkington's 'TBL: The Triple Bottom Line Model' is a model that presents the concept of sustainability in a theoretical way, taking into consideration not only the economy but also the environmental and social issues. Based on this theory, various sustainability theories are emerging and the 'business model canvas' which is discussed later has also suggested a concrete component based on the theory of the triple bottom line to the business model canvas.

2.2 Definition of Business Model

In the previous studies of business model, Timmers (1998) [6] defined business model as a model that provides the systematic structure and approach needed in the process of creating new value through new business opportunities and to providing them to customers. Amit and Zott (2001) [7] defined the business model as a model to describe the business structure, business content, and management needed to develop business opportunities and create value [8]. Teece (2010) [9] defined the business model as a model that provides data and other evidence that shows how a business is created and how it provides value to its customers. Also, he defined business model as a model that embodies an organization of a company and its financial structure [1].

To sum up the definition of the business model concept, a business model can be defined as a general direction and method of in what market, to whom, what value, and in what kind of method a company will communicate with business ideas, and of how to generate revenue [1]. In shorts, business model is a tool to help companies find new opportunities to continuously create future corporate value in order to survive in rapid technological progress and swiftly changing competitive environment.

The advantage of the Business Model Canvas is that it contains all the content that needs to be addressed in business models such as value proposition, value creation, value delivery, and value capture [1, 10]. When analyzing a business model in practice, analysis from each individual professional department is not needed [11]. Instead, each professional department uses a consistent analysis tool for business model when analyzing the elements of value creation, value proposition, value delivery, and value capturing. Through this, they can communicate and cooperate with each other freely and have an overall perspective when dealing with the business model [10].

3   Triple Layered Business Model Canvas

Joyce and Paquin (2016) [12] argued that in order for a corporation to continuously generate value, it must analyze the environmental impact from the resource's lifecycle perspective, and analyze from a social perspective that includes all stakeholders involved in the organization within the enterprise, rather than analyzing the business only from an existing economic value perspective. In short, they argued that it is necessary to organically analyze triple layers that can provide integrated analysis of economic, social and environmental aspects when analyzing the sustainability of a company. Looking more closely into the 'Triple Layered Business Model Canvas', 'the Economic Business Model Canvas' emphasizes the value of business and is the same as Osterwalder's Business Model Canvas, and the name 'economical' is used to distinguish it from other canvases.

The Environmental Business Model Canvas is proposing an analysis framework for environmental impacts from a life cycle perspective of resource recycling. The basic framework is based on the 'business model canvas', and it changes and matches each component from the lifecycle perspective of service or product. It is based on a study of Life Cycle Assessments (LCA) that formally measures the environmental impact of a product or service at all stages of the product life cycle. Analysis of the business model using LCA can help to innovate business models that generate sustainable value [12].

The Social Business Model Canvas suggests a framework for analyzing the social impact of an organization based on the management method of stakeholder related to the organization. Likewise, the basic framework is based on the 'business model canvas' and each component is modified and matched from an organizational stakeholder perspective. It extends the concept of stakeholders to include not only the typical stakeholders (employees, shareholders, communities, customers, suppliers, government agencies, etc.) but also the non-human groups such as media, the poor, terrorist groups, and natural ecosystems in order to balance the interests of stakeholders involved in the organization.

Each layer of the 'Triple Layered Business Model Canvas' presents two analytical methods to create two clear and different types of value creation. First type is the horizontal linkage analysis in which the nine elements in each economical, environmental, and social layer are combined for the analysis. The other is the vertical linkage analysis which is a method of analyzing the elements between layers by combining the elements between layers such as the social value and value proposition in the economic layer, the functional value in the environmental layer, and the value proposition in the economic layer. If the analysis is done on the relationship between business activities and components by linking the economic, environmental, and social layers, the integrate business model can be seen from a sustainability perspective [9, 12].

4   Kakao's Business Model Analysis

4.1   Kakao's Economic Business Canvas

Customer Segments

As a national messenger service, KakaoTalk is centered on general individual customers, and the customers, for whom the company provides additional services such as games and advertisements based on this, can be classified as corporate customers. Individual customers are divided into smartphone users and personal computer users, both of which are almost overlapping. When comparing domestic and overseas customers, monthly active users (MAUs) account for 43.58 million domestic customers and 6.53 million overseas customers [1]. Corporate customers are using Kakao for marketing that target the users of KakaoTalk. PlusFriend affiliates, advertisers, and, game developers are all corporate customers. Corporate customers can provide notification talk and chat talk based on PlusFriend.

Value Proposition

Kakaotalk's value offering elements can be categorized into individual customers and corporate customers. The value provided to individual customers' aspect is free chat, various linked services, portal services and so on. Free text/chat communication is a basic value. Only with a phone number, customers can chat in real-time with one-on-one chat, group chat, and easily send and share multimedia files such as photos, videos, and contacts [1]. Emotional communication is possible through emoticons and various items, and by using interesting emoticons, they can implicitly give a funny expression for the things that are hard to express in words. Through human networking based on the phone number, the company links and provides various services such as Kakao Pay, Kakao Bus, Kakao Subway, Kakao Map, and Alert Talk. In the case of Kakao Pay, it provides not only Dutch treat function but also a remittance service using KakaoTalk ID interlocking with Kakao Bank.

Channels
A channel refers to how a company communicates in order to offer value to customers, how customers evaluate the value of the services provided, and how to perform post-maintenance on the services. Basically, mobile app is provided through the smart device app market, and KakaoTalk also provides games through its own social gaming platform, 'Kakao for Game'.

Customer Relationship
In case of platform services, it is important to keep the customer in the platform. Customers use a lot of emoticons, and they can express their emotions easily, funny and witty through the shape of emoticons for the things that are difficult to chat or speak in words. Kakao offers popular emoticons unlimitedly for free or free for a limited period. The function of chat is divided into diverse chats such as general chat, open chat, and secret chat, so that the users themselves can make the service.

Revenue Streams
Kakao's revenues come from advertising, content, commerce, and mobility (Kakao Tax, Kakao Driver, etc.).

Key Resources
The core resources in mobile services are service planners and developers. The size and activity of users are the most important factors because of the nature of messenger services. Other than that, planners and developers are also important to providing good services. In order to maintain and secure users, it is necessary to have fun and attraction. The characters of Kakao (Lion, Corn, Pitch, etc.), which are becoming popular with KakaoTalk's emoticons, are becoming increasingly important as revenue streams of offline stores.

Key Activities
A company with a platform as its key resources should develop and maintain a platform while improving the customer experience to create positive network effect. The development and maintenance of a service platform based on and linked with KakaoTalk are the most important activities. In addition, content sourcing activities are important as well to improve the customer experience. Supply activities through content partners for games, music, webtoons, movies, and dramas are also important.

Key Partnership
Key partners can contribute significantly to the success of the company and affect the future growth of the company. KakaoTalk should manage its strategic partnerships with content providers such as telecommunication companies and ISPs that manage the communication network, game developers providing various contents, webtoon producers, emoticon producers, and movie and drama distributors. In addition, it is important to cooperate with government agencies in charge of privacy protection and industrial regulation. Through strategic alliances with partners, KakaoTalk can focus its core competencies on its core business.

Cost Structure
Cost Structure of KakaoTalk includes management fee to maintain IT infrastructure and labor costs to plan and develop the service [1]. In additions, fee from the supply and demand of contents and commercials are incurred.

4.2 Kakao's Environmental Business Model Canvas

Functional Value
Function value is defined as the amount of power and data used for one year for KakaoTalk service. The amount of power can be divided into server usage (electricity consumption of data center) of KakaoTalk and the amount of power consumed by KakaoTalk on smartphone of individual subscriber. Intermediate linking equipment for data transmission was excluded. From the UX point of view, accessibility and convenience have the effect of reducing power consumption by shortening the usage time [12].

Materials
The key resources for providing KakaoTalk services can be divided into IDC center, IT infrastructure resources of computer servers, and the content provided to users. For IDC center, KakaoTalk does not operate its own data center, but leases data centers from other companies through strategic alliances. In other words, KakaoTalk uses specialized vendors because external professional vendors are far more efficient in managing IDC operations that store and manage data. This can be considered as a company's choice to focus on its function to serve customers [1]. Content sourcing is an important resource for providing better value to customers based on the messenger platform. Content sourcing should be regarded as an important resource because it provides good contents through competent external content providers.

Production
In KakaoTalk, the computing power in IDC is the driving force of production. The developers and planners who develop the service, and PC for developers are all important products. The computer servers explained in 'Materials' indicate HW, and the computing power refers to all activities that use computer technology resources by running computer servers.

Supplies and Outsourcing
Businesses providing mobile or Internet services must use the wired Internet network provided by ISP and the wireless network provided by the mobile communication company [1]. However, this cannot be managed by the service provider, although it is necessary. Recently, there have been issues of network neutrality, but until now, it can be used without discrimination due to network neutrality.

Distribution

The app markets are essential because businesses need to distribute basic apps to customers and make customers to purchase content in the apps. Typical app markets are Apple's App Store and Google Play Store for Android devices [1]. In addition, there is Samsung App Store which is an app store by the manufacturer

Use phase

The number of chats, consumption by content type, and telecommunication charges have effect on service when customers use KakaoTalk. The network environment used by customers also plays an important role in the quality of service depending on Wi-Fi or LTE network [1].

End of life

Suspension of service is withdrawal of service, which may lead to a transition to a competitive service. The service may be temporarily unavailable due to replacement of smart devices or USIM changes, and if the service is withdrawn by transfer to a competitive service, the service may be permanently discontinued.

Environmental Impacts

The parts that generates most significant environmental impacts will be the computing power of the production portion, the IDC in resources, telecommunication charge in the use phase, and power consumption of the smart device. Besides YouTube, the most popular app used by smartphones is KakaoTalk (Wise App, April 2018. Total Android usage time in April: YouTube 258 million minutes, KakaoTalk 189 million minutes, Naver 12.6 billion minutes, Facebook 4 billion minutes).

Environmental Benefits

Kakaotalk uses an externally provided IDC. However, in order to use it more efficiently, it is necessary for KakaoTalk to actively consider using public cloud service. KakaoTalk should improve UX in smartphone to reduce the usage time while obtaining key information more easily and quickly.

4.3   Kakao's Social Business Model Canvas

Social Value

There is a slogan of 'Making valuable social change with users' among the corporate slogans of KakaoTalk. Activities for this purpose include 'Value with Kakao', which is a donation platform using Kakao, 'kakaomakers' that links producers and consumers, and a win-win center for mutual growth with partners. [1]

Employees

Due to recent increases of mergers and acquisitions, new workforce has been introduced [12], which consists mainly of IT developers and planners.

Governance

Kakao is a company that started out as a venture, and it makes decisions about development on a team basis. With a horizontal organizational culture based on trust, conflict, and devotion, they have eliminated the concept of rank and position from the beginning. Not only that, they have made employees to use English names among colleagues (Announced by Lee, Jae-beom, Kakao's CEO, at YouTube in 2011).

Communities

Since the headquarter is located in Jeju Island, Kakao is operating the Jeju Regional Cooperation Center, where it carries out industry-university cooperation with Jeju University to conduct engineering education innovation projects. Kakao is working with the Jeju Creative Economy and Innovation Center to run Kakako Class, supporting the Jeju Film Festival as well, and running the Kakao Win-Win Center in Seoul and Gyeonggi Province [1].

Social Culture

Kakaotalk is changing the way of communication in the society, with both positive and negative aspects. The positive aspect of KakaoTalk is that it allows free chat with others easily and conveniently. The negative aspects of KakaoTalk include KakaoTalk addiction, Kakao prison, and extension of work, which may even infringe the privacy of individuals.

Scale of Outreach

Kakaotalk is available in 16 languages in 230 countries, and is providing youth IT education and business start-up support in these regions [12].

End-Users

Kakaotalk allows users to conveniently share chat and information in a variety of ways, and enables them to express fun and rich emotions through emoticons and (#) searches, and makes them to be able to use contents easily through Kakao Page [12]. In other words, Kakao provides an easy, fun, convenient, and rich chat platform.

Social Impacts

The biggest problem of KakaoTalk is that users may be addicted to KakaoTalk, which is a new and different type of problem compared to smartphone addiction. Kakao prison and group chatting rooms for business are also causing stress to the chat.

Social Benefits

Kakaotalk has a wide range of social contribution activities. It operates a donation program by utilizing KakaoTalk platform, and contributes to nurturing future IT talents by running IT education programs for young people.

4.4 Horizontal and Vertical Linkage Analysis

The horizontal linkage analysis in each layer of the 'Triple Layered Business Model Canvas' enables one to find new business opportunities.

Figure 1 above shows the association of IDC resource among the resources used by KakaoTalk with the constituents of the environmental layer. In most cases of Internet service providers, the environmental impact by IDC is the biggest. Likewise, KakaoTalk also has the largest environmental impact in IDC resource. Thus, KakaoTalk can find opportunities to minimize the environmental impact by exploiting the opportunity to utilize the public cloud to get cheap and flexible computing resources as an alternative for environmental benefits [12]. This also shows that economic value can be obtained by reducing costs in connection with the cost structure of economic layer.

When linking social and economic business models, we can find that there is a need to introduce differential membership and point systems to attract loyal customers of the service. According to Fig. 2, the presence of a customer using other chat services while using KakaoTalk means that KakaoTalk is at risk of outflow of subscribers or probability of breakaway of subscribers from KakaoTalk [1]. To prevent this, it is necessary to utilize membership or point system. Open chat is a chat room where people with different interests can participate anonymously. Therefore, it is possible to consider using the volunteer activity room to link activity indicators with points, to encourage consumption of Kakao contents such as emoticons and webtoons, or to donate to NGOs such as World Vision or UNICEF [1]. 'Triple Layered Business Model Canvas' provides an opportunity for KakaoTalk to expand their business with

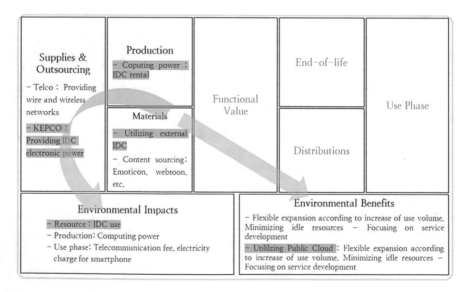

Fig. 1 Horizontal linkage analysis on KakaoTalk's economic layer

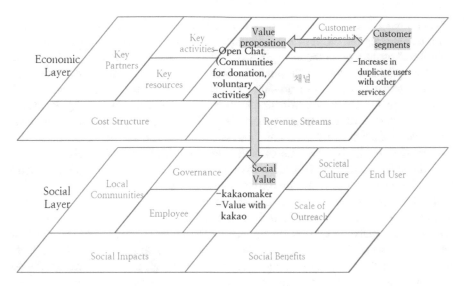

Fig. 2 Vertical consistency: linkage analysis between economic, environmental, and social layers

sustainable growth or to explore new innovations through linkage analysis of each layer.

5 Conclusion

The existing Business Model Canvas has been approaching from the viewpoint of business management, expressing the overall business structure simply and clearly, focusing on the value that the company intends to provide, and showing the business blueprint immediately. It is a convenient analytical tool that helps a company to organize its current business structure by business model components. However, there was a limitation to draw business insights to innovate the business model [1]. This is a problem that has come up in many other studies. In addition, if the innovation is achieved through change of Business Model Canvas from the existing management viewpoint, it is effective to preempt the market, but competition in the red ocean becomes excessive in a short period of time since it is easy to imitate in a situation where the IT technology and manufacturing technology is common. Moreover, the company may lose profit to its late competitors due to falling into the dilemma of First Penguin. Osterwalder, who proposed the Business Model Canvas, suggests using SWOT analysis and other strategy analysis tools, such as Blue Ocean Strategy, together with the canvas in order to complement such disadvantage.

Environmental and social issues are not problems restricted to specific regions or countries, but global problems like the problems dealt in Rio Declaration and the Paris Environmental Treaty. These issues are directly related to the existence of a company because they can play a role of trade barriers, not staying on the level of reducing the profit of the company. Therefore, it is a problem that must be solved and overcome for the enterprise to be maintained continuously [1]. However, most companies overlook environmental and social impacts because they focus only on the benefits of services or products, and they have not been able to understand how their business areas are affecting both environmentally and socially. Research on sustainability was conducted as various NGO organizations, including the UN, demanded solutions to environmental and social problems, and the 'business model canvas' analyzes the business model with nine components at three levels: economy, environment, and society. This allows you to evaluate the environmental and social value of the services provided by the company.

In this study, 'Triple Layered Business Model Canvas' was applied to KakaoTalk, a mobile internet service provider to analyze the 3 layers, i.e., the economic, environmental, and social layers, in a three dimensional and integrated manner. Through this analysis, we were able to concretely sort out the environmental and social impacts resulted from the company's business activities. Through this study, we were able to prove that companies can find new business opportunities by correlating and analyzing these results. In addition, we confirmed that there are some environmental aspects that need to be supplemented when applying the 'Triple Layered Business Model Canvas' to service companies, not manufacturing companies.

In conclusion, even if one uses the methodology of 'Triple Layered Business Model Canvas', one can identify only the individual phenomena of the economic, environmental, and social sectors if each layer is independently analyzed. However, if an integral analysis is done with the consideration of horizontal and vertical connectivity, the corporation can find opportunities for new business innovation that can create value, and can grow continuously while securing corporate sustainability.

The limitations of this study are as follows: First, the case study of KakaoTalk was a study mainly with literal materials such as previous articles, annual business reports, IR materials, media articles and internet search. The internal data could not be used because we could not make an in-depth investigation on the inside personnel of Kakao, and thus activities and opinions inside Kakao were not included in this study. It would be necessary to carry out a case study by improving the collection and research methods of data to overcome this limit.

Second, the case of only a single company, KakaoTalk, was examined intensively. The 'Triple Layered Business Model Canvas' is lacking case studies to generalize the findings of this study. In order to overcome these limitations and conduct research in a developmental direction in the future, various case analysis using the 'business model canvas' should be proceeded first. The business environment of a company varies depending on the type of industry and the situation of individual companies. Research should be continued in order to generalize and improve the analysis results of this study by conducting case studies of companies in various environments such as manufacturing, distribution, service, and start-up.

Finally, we can construct more sophisticated components of the three layers of economy, environment, and society, and to consider addition of new components by conducting research on the importance and validity of the environmental and social layers of the 'business model canvas'.

References

1. Kwak HY, Kim JS, Lee ST, Gim GY (2019) A study on the sustainable value generation of mobile messenger service using 'triple layered business model canvas'. In: 2019 20th IEEE/ACIS international conference on software engineering, artificial intelligence, networking and parallel/distributed computing (SNPD), pp 340–350, IEEE
2. Osterwalder A, Pigneur Y (2019) Business model generation: a handbook for visionaries, game changers, and challengers. John Wiley & Sons
3. Osterwalder A, Pigneur Y (2011) The birth of a business model. Time Biz Publisher
4. Jeong SJ (2013) The sustainability of consumer-oriented fashion products. Master's Thesis, Seoul National University, Seoul, Korea
5. Elkington J (1998) Partnerships from cannibals with forks: The triple bottom line of 21st-century business. Environ Qual Manage 8(1):37–51
6. Timmers P (1998) Business models for electronic markets. Electron Markets 8(2):3–8
7. Amit R, Zott C (2001) Value creation in e-business. Strateg Manag J 22(6–7):493–520
8. Amit R, Zott C (2009) Business model innovation: Creating value in times of change. Universia Bus Rev 23(1):108–121
9. Teece DJ (2010) Business models, business strategy and innovation. Long Range Plan 43(2–3):172–194
10. Lee JG (2018) Study on business model innovation of late entrant in the underwear industry: focused on effectual approach. Doctoral dissertation, Ph.D. Thesis, Catholic University, Seoul, Korea
11. Morris M, Schindehutte M, Allen J (2005) The entrepreneur's business model: toward a unified perspective. J Bus Res 58(6):726–735
12. Joyce A, Paquin RL (2016) The triple layered business model canvas: a tool to design more sustainable business models. J Clean Prod 135:1474–1486

The Effects of Product's Visual Preview on Customer Attention and Sales Using Convolution Neural Networks

Eun-tack Im, Huy Tung Phuong, Myung-suk Oh, Jun-yeob Lee, and Simon Gim

Abstract The hyper-connected society has shifted consumers' purchasing behavior from offline-oriented to online and mobile oriented, forcing consumers to make decisions under vast amount of information. In response, sellers are delivering compressed images of goods and services in order to instantly attract consumers in virtual space. This paper focused on determining whether the attributes of images for products in mobile commerce attract customers and have a significant influence on purchase decision making. In the research, regression analyzed how images provided by the sellers affect product sales page views and product sales. Using attributes such as emotion, aesthetics, and product information were extracted through Deep-CNNs model and vision API from t-shirts images sold in Singapore's leading mobile commerce application 'Shopee.' As a result of the analysis, the image information Entropy and the color-harmony representing the emotion of the image had a significant effect on the number of views. Moreover, Model, a variable indicating whether a model appears in the image, was negatively adopted.

Keywords Mobile-commerce · Customer attraction · Sale performance · Image information processing · Computer vision

E. Im (✉) · H. T. Phuong · M. Oh
Graduate School of Business Administration, Soongsil University, Seoul, South Korea
e-mail: iet030507@gmail.com

H. T. Phuong
e-mail: tungph@soongsil.ac.kr

M. Oh
e-mail: opro1226@naver.com

J. Lee
College of Economics, Sungkyunkwan University, Seoul, South Korea
e-mail: busyru98@gmail.com

S. Gim
SNS Marketing Research Institute, Soongsil University, Seoul, South Korea
e-mail: simongim93@gmail.com

© The Author(s), under exclusive license to Springer Nature Switzerland AG 2021
H. Kim and R. Lee (eds.), *Software Engineering in IoT, Big Data, Cloud and Mobile Computing*, Studies in Computational Intelligence 930,
https://doi.org/10.1007/978-3-030-64773-5_14

1 Introduction

The definition of M-commerce is the conduct of commercial transactions via mobile devices available through the Internet [1]. The growing popularity of M-commerce is because the commerce is based on devices which are wireless. Wireless device allows customers to obtain various kinds of data, such as text, picture, or even video in anytime and anyplace [2]. It means users of M-commerce could shop products they need in handier environment compared to E-commerce environment. In mobile environment, it is found that a mobile platform encourages the customers to purchase more, compared to selling solely on web environment [3]. Thus, it is important to research M-commerce which is already taking a large portion for articles and would be more flourishing in near future [4].

However, there are only few researches conducted for visual representation in M-commerce which might be more dependent on image due to being relatively smaller than e-commerce. Cyr et al. [5] states that visual design of mobile domain significantly related to user's positive experience could influence user loyalty toward the domain. Furthermore, Li and Yeh [6] argues that design of mobile website has an explanatory effect on customer trust. Those studies show that visual aspects are essential to induce user to have positive experience in M-commerce. Related to visual aspects, product images might be important for customers experience. Concrete picture used in E-commerce has been shown to facilitates behavioral intentions by stimulating positive emotion than less concrete picture [7]. In addition, in E-commerce retail, visual stimuli influence customer' imagery processing [8].

This paper extracted images of men's t-shirts sold in Shopee, an m-commerce that has the most users in Southeast Asia, and the number of sales page views and product sales. The attributes of the image were derived from the crawled image extracted using deep-CNNs model, vision API, and entropy calculation. Regression analysis was conducted to determine which characteristics customers access and purchase products through the organized data.

2 Theoretical Background

2.1 Relationship Between Store Attributes and Customer Attention in Online Retailing

Researches have shown that attributes of offline vender, such as employees, the display of products, and odor, influence consumer behavior [9, 10]. Since E-commerce and M-commerce have introduced in various market, the trend of research for offline shop have been switched to online. Eroglu et al. [11] find out high task-relevant cues (e.g., product description, price, and photography) contribute customers practical desire but low task-relevant cues (e.g., font and sound) are relevant to hedonic desire. Beyond the research that show the quality of website, consumer behavior, favor are correlated in positive way [12, 13], Wu et al. [14], based on The Theory of reasoned action (TRA), states that even actual purchase behavior could be increased when they have impressive experience in online mall.

However, even though the information, offered from online vendor, could enhance customer satisfaction, there is a drawback. According to the theory of information overload, if customers suffer from a large amount of information which overload their perception tolerance, the information detract attention of customer. This could negatively influence purchase decisions of customers [15]. Furthermore, Xia et al. [16] research shows that fulsome visual clue could burden client memory [17], because of time and perception limit that consumers have [18]. Therefore, they emphasize to research the relationship between the attributes of product images and product sales based on information system.

2.2 Customer's Visual Information Processing on the Web and Product Visual Preview

In online environment, consumer cannot evaluate products directly, and online venders provide various indirect clues to customer in order to stimulate their desires. Preview can be the typical example of these clues, defined as the information that sellers offer to customers to indirectly experience beforehand. Such information includes photograph of products, even for three-dimensional photo. MacInnis and Linda [19] revealed that since images last in memory longer than verbal information, they could affect more on customers' behavior by provoking sensory experience. In online shopping, because consumers tend to reduce their uncertainty of products [20], images are essential channel to providing useful information to visitors. Especially, characteristics of photograph, such as the size and proper representative of details could enhance purchase intention of consumers [21].

Yoo and Kim [7] states that solid image of products is key to online vendors because it is effective to stimulate virtual experience of customers, which is directly related to behavior intention. Junaini and Sidi [22] properly used color of product images help the web page to be attractive, which could appeal to young population to visit the web site. Moreover, improved display could boost sales by offering decent experience of online mall.

This research aims to find out which attributes of product images offered by M-commerce sellers are related to customers' intention of purchases and even actual purchases. Since the display of M-commerce are generally smaller than E-commerce, it could be expected that the effects of images might be different from the research with E-commerce setting.

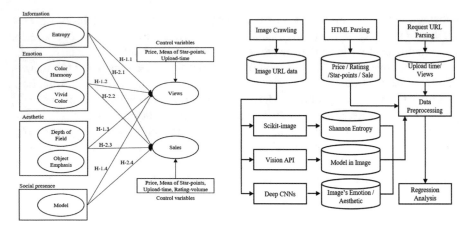

Fig. 1 Research model and date analysis process

3 Research Design

Based on previous studies in the literature [16, 21, 23–26], this study examines four groups of attributes that product images may contain. The four groups are information, emotion, aesthetic, and social presence. Information embedded in the image can be measured simply by Shannon's Entropy [21, 27]. To measure the emotion of images, we employed two types color-related measures: the degree of vividness and the degree of harmonization of color. To measure the aesthetic attribute, we examined the level of depth of the field and object emphasis. For the social presence measurement, research was based on whether the photo contains human models or not. For brevity, these independent variables were named as follows: "Entropy", "Color Harmony", "Vivid Color", "Depth of Field", "Object Emphasis", "Model".

The two dependent variables to evaluate the level of customer acquisition and sales ability are the number of views (views) and number of sales (sales) respectively, as shown in Fig. 1. We will explain more about the process in the chapter 4—"Empirical Data Analysis". And Table 1 is the operational definition of the variables used.

4 Empirical Data Analysis

4.1 Data Collection

Founded in 2015, Shopee is the largest mobile commerce platform in Southeast Asia and Taiwan. Shopee was the top Shopping App by Monthly Active Users, Total

Table 1 Operational definition

	Variable	Definition	Reference
Control Variables	Upload time	Duration of exposure of product, by counting milliseconds since epoch when the product initially uploaded	
	Mean of Star-points	Mean rating (1 ~ 5) of product, contributed by clients who actually buy the products	
	Rating Volume	The amount of rating records, classified as 4 groups from the fewest group to the largest group	
	Price	Price of product, classified as 4 groups, ordered from the cheapest group to the most expensive group	
Image's Information	Model	Dummy variable represents existence of model in product photograph 1: exist, 0: not exist	[16, 21, 23–25]
	Entropy	Complexity of images. The larger means the more information included	[21, 26]
Image's Emotion	Color Harmony	Harmony of colors. The large number states colors in image are harmonious	[21, 28, 29]
	Vivid Color	Diversity of colors. The more colors used in image, the variable approaches 1	[21, 28, 29]
Image's Aesthetics	Depth of Field	It represents depth of image. The number close to 1 means the image include more perspective by blur the background	[21, 28, 29]
	Object Emphasis	Degree of emphasizing an object compared to the other objects in image	[21, 28, 29]
Dependent variables	Views	View counts of the product counted from last 30 days, transforms as natural logarithm	[21]
	Sales	Cumulative sales of the product, transforms as natural logarithm	[16]

Downloads, and Websites with the Most Visits. Since Shopee is a mobile application-oriented m-commerce company, it was judged as a suitable research target for the study on the effect of image on customer attraction and sales. Based on the page where sales products are displayed to consumers, images of 500 products and data necessary for research in the product sales page of t-shirts were collected through crawling for 2 days from July 15 to 16, 2020. The first product uploaded in the 500 products collected for the data was November 22, 2016, and the most recent product was uploaded on June 14, 2020.

To achieve data the research needed, two methodologies were used. Firstly, HTML parsing, the general way to extract data from website, is conducted to get image url, mean of star-points, price rating volume, and sales. Secondly, in order to withdraw registration date and view count, Shopee request URL were used. In detail, for fetching data to bring up on screen, the web server must request call to server where the data is stored to get data. By using identification number of product and vender, obtained from URL, it is allowed to access data, as JSON format, which get requested. From the data, the study used 'Upload time' and 'Views'. 'Upload time' generally refers to the creation time. However, to be corroborated, the possibility of contradiction is checked by comparing first rating date and 'upload time'. Furthermore, 'Views' are used in sellers' web page, to inform how many people actually click the product. According to Shopee FAQ, it is stated that view count is measured only for 30 days.

4.2 Image Attributes Extraction

4.2.1 Image Attributes Extraction Using Vision API

Python programing and Google Vision API were used for human recognition in product images. First, we ran multiple objects detection function of the API to get information (list of objects and bounding location) data for the objects in the image. We just focused on "person" objects and marked the image that has "person" with "human model available" label. A demo of using API can be seen in the Fig. 2.

4.2.2 Image's Attributes Using CNNs

We used Deep Convolutional Neural Networks to solve this task. The model was proposed by Kong, S. et al. [28], improved by Malu, G. et al. [29] can be found in Fig. 3. A user in Github.com, kevinlu1211 [30], has fine-tuned and implied the model with Pytorch framework, an open source machine learning library based on the Torch library. The final model was recognized for its high efficiency and accuracy in evaluating the level of these attributes: "Color Harmony", "Vivid Color", "Depth of Field", "Object Emphasis" [29]. An example using the algorithm proposed in this paper can be found in Fig. 4.

4.2.3 Image Entropy

The entropy or average information of an image, first introduced in information theory by Shannon, C. E. [27], is a measure of the degree of randomness in the image [31].

Fig. 2 Demo for human recognition using Google Vision API

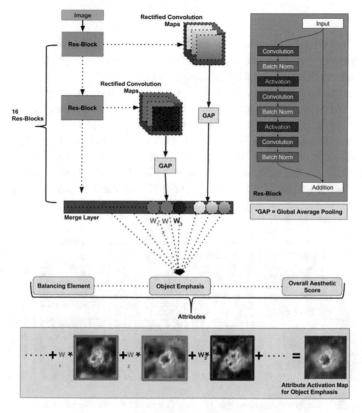

Fig. 3 Approach for generating attributes of images by Malu et al. [29]

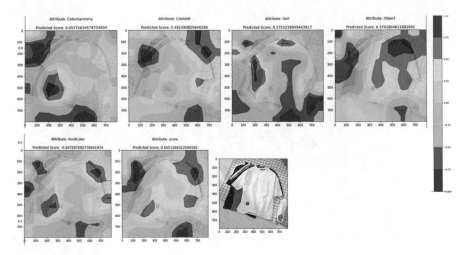

Fig. 4 Sample output extracted using Malu's model (The attributes except for "overall score" are normalized to the range of [−1, 1])

It can be determined approximately from the histogram of the image which shows the different grey level probabilities in the image. Entropy is a quantitative measure of the information transmitted by the image [32].

To calculate entropy, the order of the pixels is ignored and only the distribution P of the RGB values of the pixels are used. Color images are often converted to a grey-scale image by averaging the RGB values [21]. In our empirical analysis, we applied a measure module of Skicit-image, an open-source image processing library for the Python programming language, to calculate the Shannon entropy of images. Figure 5. is shows examples of the images which have the lowest and the highest predicted values extracted by using Deep-CNNs, Scikit-image and Google Vision API.

Fig. 5 Sample of Shopee product images labeling (Top row: High rated images, bottom row: Low rated image)

Table 2 Descriptive statistics of the data

Variable	Frequency	Min	Max	Mean
Product	468			
Upload-time	468	1479796305	1592150559	1565252758
Upload-date	468	2016-11-22	2020-6-14	2019-8-8 3:25
Sales	468	8	6900	192.4017
Views	468	11	16128	842.3782
Price	468	2.32	135	11.2329
Mean of star-points	468	0	5	4.6028
Ratings volume	468	1	1700	70.4038
Model(0)	242			
Model(1)	226			
Entropy	468	2.8760	13.6007	8.7321
Color Harmony	468	−0.0768	0.6894	0.3682
Vivid Color	468	−0.3567	0.8217	0.1726
Depth of field	468	−0.2193	0.7234	0.1963
Object Emphasis	468	−0.4372	0.8987	0.6011

* Control variables (price, rating volume): Dividing the degree into quartile
* N/A, null values deleted

4.3   Data Preprocessing and Coding

Of the data for 500 extracted products, 32 products without sales or views were deleted. Descriptive statistics of the final 468 products analyzed can be found on Table 2.

4.4   Regression Analysis

4.4.1   Effects of Image Attributes on Customers Attraction

When a consumer searches for a product category in the Shopee, information on images, prices, and average ratings of various products can be recognized, and a detailed sales page of the product to be purchased can be accessed. The purpose of this paper is to find out the effect of product image attributes on product sales page views. Product sales, upload time, star rating, and price are used as control variables. The regression analysis results are summarized in Table 3.

Model ($\beta = -0.232$, p $= 0.034$) and Entropy($\beta = 0.068$, p $= 0.005$) in image's informational attributes were all adopted. Among them, the model was expected to have a positive effect on the number of views, but had a significant negative effect. This means that if the image contains only the products, not models, the product

Table 3 Results of Regression Analysis (effects on views)

Hypothesis	Independent variable	Estimate (β)	t-value	Supported/not supported
Control	Upload-time	−0.075	−1.733*	
Control	Mean of star-points	0.403	5.445***	
Control	Price	0.538	3.136**	
H-1-1.1	Model	−0.232	−2.121**	Supported
H-1-1.2	Entropy	0.068	2.791**	Supported
H-1.2.1	Color Harmony	1.361	2.870**	Supported
H-1.2.2	Vivid Color	−0.238	−1.018	Not supported
H-1.3.1	Depth of Field	−0.477	−1.117	Not supported
H-1.3.2	Object Emphasis	−0.483	−1.646	Not supported

$* = p < 0.05, ** = p < 0.01, *** = p < 0.001$

detail page is accessed. Only color harmony ($\beta = 1.361, p = 0.004$) was adopted in the Emotional attributes, and all were rejected in the Aesthetics attributes.

4.4.2 Effects of Image Attributes on Customers Attraction

Customers' purchasing behavior is done on the product detail page, and before purchasing the product, customers can check the product's material, rating, and text review. Previous studies have shown that the mean of rating-points and ratings volume are important influence on the behavior of customers when purchase products. Therefore, regression analysis was performed using the number of evaluations added to the price, average rating, and uploaded time. Later, it was used as control variables. The analysis results are summarized in Table 4. According to the results, only color

Table 4 Results of hypotheses tests (effects on sales)

Hypothesis	Independent variable	Estimate (β)	t-value	Supported/not supported
Control	Upload-time	−0.551	−13.881***	
Control	Mean of star-points	0.249	3.699***	
Control	Rating volume	2.453	8.836***	
Control	Price	−0.286	−1.832***	
H-2-1.1	Model	−0.136	−1.366*	Not supported
H-2-1.2	Entropy	0.034	1.523	Not supported
H-2.2.1	Color Harmony	0.782	1.810*	Supported
H-2.2.2	Vivid Color	−0.309	−1.450	Not supported
H-2.3.1	Depth of Field	−0.172	−.442	Not supported
H-2.3.2	Object Emphasis	0.186	.696	Not supported

$* = p < 0.05, ** = p < 0.01, *** = p < 0.001$

harmony in emotional attributes of image was adopted ($\beta = 0.782$, p $= 0.004$).

5 Conclusions

5.1 Research Summary and Implication

Images are provided by the seller to the customer before the purchase, causing the consumer to remember the product. In addition, product images are important because when customers need to make uncertain decisions about product purchases in online virtual space, visual experience of the product can help them identify the attributes of the product. This study analyzed the impact of product image attributes that attract customers in mobile commerce where purchasing decisions should be made on small screens of mobile devices. In the hypothesis of the customer's access to the product sales page, 'Entropy', which indicates the amount of information in the product, had a positive effect, and in social presence, 'Model' had a negative effect. This shows that consumers prefer having a specific image that contains information about a product rather than an image that contains a model that affects the customer's intention to purchase. This is same as previous research that suggests that the elaborately captured image is effective in promoting the virtual product experience claimed by Yoo and Kim [7]. "Color Harmony" was adopted and "Vivid Color" was rejected, which is more important to balance color harmony than to use various colors. All aesthetic factors in the image have been rejected, meaning that information and feelings are more important to the consumer than how the photograph is taken. This paper is meaningful in that the product image attributes of M-commerce field are divided into information, sensibility, aesthetic, and social factors to examine the influence relation.

5.2 Limitations

This paper did not properly explain the direct impact of images on sales. There is no cost risk when the customer is lured and motivated to explore the goods. However, because there is a cost risk at the stage of purchase, consumers will make purchasing decisions through information processing using more diverse information pages. Therefore, it is necessary to study the direct and indirect effects of images on purchasing decisions through a deeper understanding of the consumer's information processing process in regard to the effects of images on sales and the important factors influencing consumers' purchasing decisions in online space. Moreover, it was confirmed that the characteristics of consumer reviews, which are being conducted in various studies and were used as control variables in this study, had great influence on purchasing decision making. Therefore, in future research, it is necessary to analyze the effect of the attributes of reviews and images, which are information that customers can recognize when making decisions in mobile commerce, on customer behavior.

References

1. Omonedo P, Bocij P (2014) E-commerce versus M-commerce: where is the dividing line. World Acad Sci Eng Technol Int J Soc Behav Educ Econ Bus Ind Eng 8(11):3488–3493
2. Jahanshahi AA, Mirzaie A, Asadollahi A (2011) Mobile commerce beyond electronic commerce: issue and challenges. Asian J Bus Manag Sci 1(2):119–129
3. Huang L, Lu X, Ba S (2016) An empirical study of the cross-channel effects between web and mobile shopping channels. Inf Manag 53(2):265–278
4. Ngai EW, Gunasekaran A (2007) A review for mobile commerce research and applications. Decis Support Syst 43(1):3–15
5. Cyr D, Head M, Ivanov A (2006) Design aesthetics leading to m-loyalty in mobile commerce. Inf Manag 43(8):950–963
6. Li YM, Yeh YS (2010) Increasing trust in mobile commerce through design aesthetics. Comput Hum Behav 26(4):673–684
7. Yoo J, Kim M (2014) The effects of online product presentation on consumer responses: a mental imagery perspective. J Bus Res 67(11):2464–2472
8. Kim M (2019) Digital product presentation, information processing, need for cognition and behavioral intent in digital commerce. J Retail Consum Serv 50:362–370
9. Baltas G, Papastathopoulou P (2003) Shopper characteristics, product and store choice criteria: a survey in the Greek grocery sector. Int J Retail Distrib Manag
10. Carpenter JM, Moore M (2006) Consumer demographics, store attributes, and retail format choice in the US grocery market. Int J Retail Distrib Manag
11. Eroglu SA, Machleit KA, Davis LM (2003) Empirical testing of a model of online store atmospherics and shopper responses. Psychol Mark 20(2):139–150
12. Lynch PD, Kent RJ, Srinivasan SS (2001) The global internet shopper: evidence from shopping tasks in twelve countries. J Advert Res 41(3):15–23
13. Kim J, Fiore AM, Lee HH (2007) Influences of online store perception, shopping enjoyment, and shopping involvement on consumer patronage behavior towards an online retailer. J Retail Consum Serv 14(2):95–107
14. Wu WY, Lee CL, Fu CS, Wang HC (2014) How can online store layout design and atmosphere influence consumer shopping intention on a website? Int J Retail Distrib Manag
15. Chen YC, Shang RA, Kao CY (2009) The effects of information overload on consumers' subjective state towards buying decision in the internet shopping environment. Electron Commer Res Appl 8(1):48–58
16. Xia H, Pan X, Zhou Y, Zhang ZJ (2020) Creating the best first impression: designing online product photos to increase sales. Decis Support Syst 131:113235
17. Jiang Z, Benbasat I (2007) The effects of presentation formats and task complexity on online consumers' product understanding. Mis Quarterly 475–500
18. Koufaris M (2002) Applying the technology acceptance model and flow theory to online consumer behavior. Inform Syst Res 13(2):205–223
19. MacInnis DJ, Price LL (1987) The role of imagery in information processing: review and extensions. J Consum Res 13(4):473–491
20. Laroche M, Yang Z, McDougall GH, Bergeron J (2005) Internet versus bricks-and-mortar retailers: An investigation into intangibility and its consequences. J Retail 81(4):251–267
21. Li X, Wang M, Chen Y (2014, June) The impact of product photo on online consumer purchase intention: an image-processing enabled empirical study. In PACIS (p 325)

22. Junaini SN, Sidi J (2005, December) Improving product display of the e-commerce website through aesthetics, attractiveness and interactivity. In CITA (pp 23–27)
23. Moshagen M, Thielsch MT (2010) Facets of visual aesthetics. Int J Hum Comput Stud 68(10):689–709
24. Bauerly M, Liu Y (2006) Computational modeling and experimental investigation of effects of compositional elements on interface and design aesthetics. Int J Hum Comput Stud 64(8):670–682
25. Cyr D, Head M, Larios H, Pan B (2009) Exploring human images in website design: a multi-method approach. MIS Quarterly 539–566
26. Wang M, Li X, Chau PY (2016, December) The impact of photo aesthetics on online consumer shopping behavior: an image processing-enabled empirical study. In 37th international conference on information systems, ICIS 2016. Association for Information Systems
27. Shannon CE (1948) A mathematical theory of communication. Bell Syst Tech J 27(3):379–423
28. Kong S, Shen X, Lin Z, Mech R, Fowlkes C (2016, October) Photo aesthetics ranking network with attributes and content adaptation. In European conference on computer vision (pp 662–679). Springer, Cham
29. Malu G, Bapi RS, Indurkhya B (2017) Learning photography aesthetics with deep cnns. arXiv preprint arXiv:1707.03981
30. kevinlu1211. Deep photo aesthetics (2018) GitHub repository: https://github.com/kevinlu1211/deep-photo-aesthetics
31. Thum C (1984) Measurement of the entropy of an image with application to image focusing. Optica Acta: Int J Optics 31(2):203–211
32. Tsai DY, Lee Y, Matsuyama E (2008) Information entropy measure for evaluation of image quality. J Digit Imaging 21(3):338–347

Implementation of Cloud Monitoring System Based on Open Source Monitoring Solution

Eunjung Jo and Hyunjae Yoo

Abstract In this paper, a cloud monitoring system based on an open-source was designed and implemented by subdividing it into CPU, memory, storage, and network parts. Linux and KVM operating systems were used to implement the designed system, and Zabbix and Grafana were used as open-source monitoring solutions. Also, an open source-based monitoring system was implemented to facilitate scalability, such as upgrading and adding monitoring functions, and a smooth and flexible monitoring system compared to commercial products provided by Cloud Service Provider. To review the cloud monitoring system designed whether it works. It was implemented on the cloud, and it was confirmed that CPU utilization, memory utilization, volume, and network resources are generally monitored in real-time.

Keywords Cloud · Open platform · Resource management · Service vendor · Open-source monitoring

1 Introduction

As the I.T., the environment rapidly changes to virtualization, the number and types of various resources on the cloud that must be managed per I.T. operators have increased exponentially, and interest inefficient cloud monitoring is rising. Monitoring cloud resources is a necessary technology, one of the various studies involved in the transition to the cloud for the efficient operation of I.T. resources. In a cloud system, information on countless nodes, V.M.s must be continuously collected, and monitoring of each resource (Computer, Storage, Network, Service, etc.) for the corresponding of this subject needs Node and V.M.s monitoring [1]. This includes efficient allocation of resources, reduced downtime through rapid failure overcoming, detection of resource usage changes, and relocation or integration of V.M.s to the right place, for moreover, stability, scalability, and elasticity will provide an increase

E. Jo (✉) · H. Yoo
Soongsil University, Seoul, Korea
e-mail: eunjung.jo1@gmail.com

© The Author(s), under exclusive license to Springer Nature Switzerland AG 2021   181
H. Kim and R. Lee (eds.), *Software Engineering in IoT, Big Data, Cloud and Mobile Computing*, Studies in Computational Intelligence 930,
https://doi.org/10.1007/978-3-030-64773-5_15

of [1]. However, research on cloud monitoring is slow. In a cloud environment, monitoring targets are often erratic and changing; they are large in scale and often have different platforms [1]. Therefore, a monitoring system in a cloud environment should support heterogeneous communication, respond flexibly to state changes, and have little performance change according to scale [1].

In this paper, we propose an open source-based cloud monitoring system and propose a method to implement an easy and flexible monitoring system. Section 2 describes related research, while in Sect. 3 describes the open-source cloud solutions and monitoring platforms. Section 4 describes the design of a cloud monitoring system, and in Sect. 5, describes the information on the system implementation process and the advantages of building an open source-based monitoring system. Section 6 describes the conclusion.

2 Related Research

Various studies are underway on monitoring systems in cloud computing environments. Gartner, a leading U.S. I.T. research firm, has been counting cloud computing as a strategic technology for years, and many other companies are using cloud computing to overcome the limitations of computer resources and realize previously impossible technologies [1]. Besides, the areas used to utilize distributed computing before cloud computing have been integrated into one through cloud computing, making it convenient and useful to use [1]. In addition to public cloud systems utilizing external resources, private cloud systems that build and use cloud systems internally are also being studied as open-source platforms such as Openstack, Eucalyptus, and CloudStack [1]. These open-source platforms have been continuously developed and developed, and Openstack recently began supporting Hadoop for big data analysis by adding a component called Sahara [1].The open-source platform, which is easy to handle and supports high performance, has lowered entry barriers, making it easy for those new to implement the cloud environment [1]. Most cloud platforms are modular, so studies suggest efficient and convenient structures by improving or integrating performance through links between different platforms [1]. Monitoring systems are one of the essential items for developed and implemented IaaS cloud systems. Cloud monitoring systems manage multiple nodes and exchange information on virtualization on those nodes to help plan ahead for service-level agreements (SLAs). Also, research and development of monitoring systems are needed to collect logs, troubleshoot problems, analyze performance, and block threats, which are the primary purposes of monitoring [1]. Cloud monitoring requires several conditions to be met. Specific features of cloud computing include On-Demand, which NIST pays for, Broad network access to allow access over the network, resource pooling to add resources as needed, Rapid Elasticity, which quickly adjusts usage according to the users will, and the Meshed Service, which measures the type or amount of services used. Cloud monitoring systems need to be designed with these characteristics in mind [1].

These studies are monitored based on the Web Interface to be independent of the platform. However, as time goes by, computer resources grow in size, and network resources that must be used continuously are growing relatively slowly [1]. Existing monitoring systems using the Web Interface as a means of transport for monitoring these cloud systems result in reduced performance due to the growing size of the systems, and inefficient use of network resources, including headers or metadata that are unnecessary for data transfer [2–4]. Studies are also underway to compare performance by proposing new transmission methods, such as REST and DDS, to reduce the consumption value of monitoring data, or by proposing a corresponding architecture for areas where performance reduction may occur [5–7].

3 Cloud Monitoring Systems

3.1 Cloud Monitoring

Cloud monitoring monitors the health of cloud resources operating the service and notifies people of the status of failures when the set threshold is exceeded. This enables rapid response to failures and plans to improve resource capacity and performance, ultimately ensuring continuity of service provided by users. Cloud monitoring services provide system-related indicators such as CPU Utilization, memory, and storage utilization. By setting up events by thresholds and utilizing specific monitoring functions such as functions and performance by computing resources, effective and rapid fault analysis is possible to respond quickly. Collects monitoring performance information for detail items and sets event alerts related to detail items. This makes more ability to monitor details [8].

3.2 Open Source Monitoring Solution

Cloud systems use monitoring tools that support the monitoring of distributed or virtualized environments. Although various open-source monitoring solutions have been studied and used, they can only be helpful if the optimal product is selected and used well depending on the purpose of introduction, but if misused, it will be cumbersome and confusing. Outline and characterize monitoring products that have received attention so far and those that are expected to develop in the future [9].

3.2.1 PROMETHEUS Monitoring Solution

Prometheus is a time series database developed around SoundCloud, a Berlin-based music distribution platform provider, and developed in the Go language. If Docker is installed on the monitoring node, both servers, clients, and alert managers can be installed as containers. The Prometheus installed on the monitoring node is connected to the Web API and references each Node directly from the browser. Data is stored in LevelDB, a key-value type NoSQL product, and a client library written in Go/Java/Ruby/Python can be downloaded from GitHub, used widely for

Docker monitoring purposes. However, there is no technical support from the vendor, and only the community should consider that.

3.2.2   FLUENTD + INFLUXDB + GRAFANA

Fluentd was developed as a log collection management tool developed by Treasure Data in C language and Ruby. When operating multiple servers, the logs stacked on each server can be integrated simply and can be handled in real-time streaming rather than Batch. InfluxDB is an open-source time-series database that stores data through RESTful API. Time series analysis is possible in SQL query language. Grafana is an open-source dashboard that connects directly to various data sources and visualizes them in real-time without programming them. Graphic, Elasticsearch, InfluxDB, and Zabbix support various data sources, and the data is inquired from the data source and expressed in graphs. Developed in Go language, data registered in D.B. can be processed, and graph generated. D.B. can also use RDB products such as MySQL or PostgreSQL, in addition to InfluxDB.

3.2.3   MUNIN

MUNIN is a product that focuses on monitoring system resource usage. It consists of a central administration server (Master) and a client agent (Node) developed as C and Perl. Master supports the Devian/Red Hat family of Linux and FreeBSD / Solaris environments. Node works with AIX/HP-UX/Mac OS X/Windows. Resource information, such as CPU/memory/disk/network, is retrieved and graphed by running the SNMP plug-in, which is installed by default by client agents installed on the Node. Monitoring of popular services such as Httpd and RDB is provided as an essential plug-in, and plug-in can be developed to perform monitoring that is not provided. However, if the number of nodes increases, the response is reported to deteriorate, so adjustments such as settings and changes in related software are needed.

3.2.4   ZABBIX

Zabbix is a centralized monitoring tool that places additional components, such as Zabbix Server and Web Server, on the central system and installs agents on the objects to be monitored to extract monitoring data [1]. Monitored data is provided through the Web Interface or Dashboard. It can be managed through the Dashboard and customized as needed [1]. Customization allows the content of the extracted monitoring information or the settings for a specific event [1]. Zabbix supports distributed monitoring methods. As cloud systems scale, the elements that makeup and leverage them increase, increasing the complexity and scale of the systems [1]. In this paper, we intend to design and implement monitoring systems that can satisfy the characteristics of cloud systems, such as Scalability and Elasticity, using Zabbix and Grafana [1].

4 Designing Cloud Monitoring Systems

4.1 Definition of Monitoring Targets

In this paper, I aim to achieve the cloud service availability goal by comprehensively monitoring the service status of significant cloud infrastructures such as CPU, memory, disk, and network in the cloud environment. First of all, the cloud infrastructure monitoring targets and message content and recipients are defined based on items, thresholds, and notification methods. The group of candidates to be monitored first monitors events such as active or stop service status of resources, utilization of services such as CPU, Memory, Storage, and service failures. The targets and items are common, but the thresholds for each target are optimized slightly differently. Table 1 shows the cloud-based monitoring items.

Table 1 Cloud-based monitoring topics

Segment	Items	Thresholds	Tool
Compute	CPU Utilization	3 min., average over 80%	Zabbix Agent
	Used Memory	3 min., average over 80%	Zabbix Agent
	DiskSpace Utilization	1 min., less than 20% of remaining space	Zabbix Agent
	Ping Loss	1 min., at least 1 time	Zabbix Agent
	Zabbix Agent Unreachable	5 min., at least 1 time	Zabbix Agent
DBCS	CPU Utilization	3 min., average over 80%	Zabbix Agent
	Used Memory	3 min., average over 80%	Zabbix Agent
	DiskSpace Utilization	1 min., less than 20% of remaining space	Zabbix Agent
	Ping Loss	1 min., at least 1 time	Zabbix Agent
	Zabbix Agent Unreachable	5 min., at least 1 time	Zabbix Agent
	Connections	1 min., Cumulative 100+	Zabbix Agent(ODBC)
Load balance	Unhealthy Backend	1 min., at least 1 time	OCI CLI
	HTTP 4XX response	5 min., Cumulative 100+	OCI CLI
	HTTP 5XX response	5 min., Cumulative 50+	OCI CLI
	Response time (HTTP ONLY)	5 min., up to 10 or more	OCI CLI
VPN	VPN TunnelState	5 min., at least 1 time	OCI CLI
Fast connect	Connection State	5 min., up to 1 or less	OCI CLI
Application	App Port Check	1 min., at least 1 time	Zabbix Agent
	Web URL Check	1 min., at least 1 time	Zabbix Agent

4.2 Integrated Operations Monitoring Screen Design

Aggregates the state of the entire instance or the computer instance by service group in a specified interval. Top 5 of display's history, sorted in order of CPU, memory, and storage utilization among all computer instances. Displays the service's operational status so that the Web service for each business can check that the service is up and running. In addition, to check the network status, the status of Load Balancer, Fast Connect, VPN Tunnel, Etc., is displayed. Figure 1 shows the Integrated Operations Monitoring Dashboard screen.

To understand of facility's resources to scale up or scale out when needed in the future and to identify the failures due to lack of resources proactively, the inventory status by cloud computer resource type reviewed and expressed, as shown in Fig. 2.

It is configured to check the status and details of events so that alarms can be checked according to the threshold values of cloud-based monitoring items. When an event occurs, the alarm details are shared with the operating personnel through slack or e-mail. This allows for the prediction of disability factors before the occurrence of a failure and proactive measures to be taken. In the event of a failure, it is required to detect and disseminate the operators' failure so that they can take prompt action (Fig. 3).

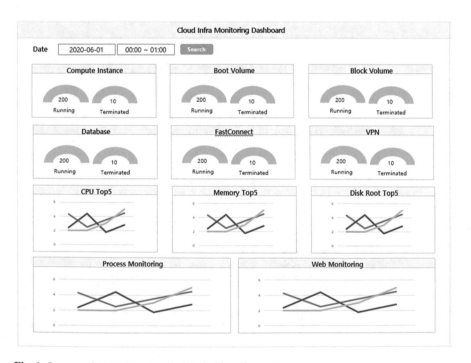

Fig. 1 Integrated operations monitoring dashboard screen

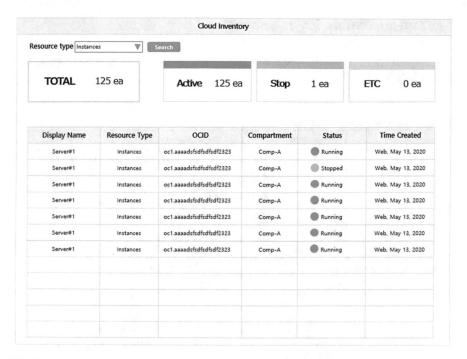

Fig. 2 Inventory status by cloud computer resource type screen

Fig. 3 Event status screen

5 Implementation Environment

The implementation environment of monitoring systems for cloud computing is shown in Table 2 and Fig. 4.

Table 2 Implementation Environment

division	Component		Emphysema
Hardware	VM	CPU	2.0 GHz Intel Xeon Platinum 8167 M
		Memory	60 GB
		Storage	500 GB
Software	OS	Linux	Oracle Linux
	Database		Maria DB
	Hypervisor		Oracle Linux Kernel-based Virtual Machine
Monitoring solutions	ZABBIX		4.0 LTS version
	Grafana		V6.7.1

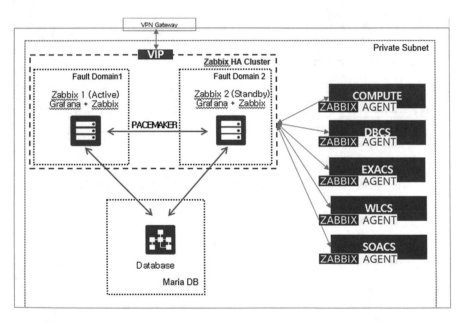

Fig. 4 System architecture

5.1 Advantages of Implementing Open Source Monitoring Services

Monitoring service to ensure high availability as well as easily and flexibly building monitoring of resources (virtual machines, storage, etc.) and cloud service performance and availability for corporate users who are considering using cloud services was composed of HA Cluster. Zabbix, built in this study, is a tool that can collect and display millions of monitoring metrics from tens of thousands of computing resources in real-time. By providing detailed and diverse functions such as inventory, reporting, SLA calculation, Web to enable the on-premise level of resource monitoring.

6 Conclusion

Cloud monitoring tools provided by cloud service vendors are vendor dependent and feature limited. To perform an efficient monitoring service, research is needed on systems that are not vendor dependent and can be analyzed variously. In this paper, an open source-based cloud resource monitoring system was designed and implemented by dividing into CPU, Memory, Storage, and Network sectors and modularizing each one. The system's structure was modularized for each possible function, considering scalability such as function upgrade, and the monitoring server was composed of an H.A. cluster for high availability. In the future, I will utilize A.I. in the cloud monitoring research sector to extract appropriate thresholds for monitoring by resource and to maintain a certain level of monitoring and contribute to the continued development of cloud computing services through research that responds quickly and actively in the event of a failure.

References

1. Kim G (2016) IaaS cloud monitoring system using message queuing = message queuing monitoring system for IaaS cloud publication. Hanyang University: Department of Computer and Software
2. BBelqasmi F, Singh J, Bani Melhem SY, Glitho RH (2012) SOAP-based vs. RESTful web services: a case study for multimedia conferencing. 2012 IEEE Internet Comput 16(4):54–63
3. Gazis V, Gortz M, Huber M, Leonardi A, Mathioudakis K, Wiesmaier A, Zeiger F, Vasilomanolakis E (2015) A survey of technologies for the internet of things. Wireless Commun Mobile Comput Conf (IWCMC)
4. Alkhawaja AR, Ferreira LL, Albano M (2012) Message oriented middleware with QoS Support for SmartGrids. Hurry-TR-120709
5. Katsaros G, Kubert R, Gallizo G (2011) Building a service-oriented monitoring framework with REST and nagios. In: 2011 IEEE international conference on service computing (SCC), pp 426–431

6. An KH, Pradhan S, Caglar F, Gokhale A (2012) A publish/subscribe middleware for dependable and real-time resource monitoring in the cloud. In: SDMCMM'12 Proceedings of the workshop on secure and dependable middleware for cloud monitoring and management
7. Corradi A, Foschini L, Povedano-Molina J, Lopez-Soler JM (2012) DDS-enabled Cloud management support for fast task offloading. In: 2012 IEEE symposium on computers and communication (ISCC), 1–4 July, pp 67–74
8. Monitoring: https://www.ncloud.com/product/management/monitoring
9. opennaru: https://www.opennaru.com/apm/open-source-monitoring-tool/

A New Normal of Lifelong Education According to the Artificial Intelligence and EduTech Industry Trends and the Spread of the Untact Trend

Cheong-Jae Lee and Seong-Woo Choi

Abstract The edutech-based lifelong education system is the most pressing alternative for democratizing education and improving the quality of education by resolving polarization in education. Lifelong education transforms the educational purpose of 'continue the individual's physical and personal maturity and social, economic and cultural growth and development throughout life' into 'educational infrastructure to foster democratic citizens who lead the future society' shall. This paper presents a New Normal of lifelong education as a lifelong education classification system innovation, self-directed learning ability cultivation, future-oriented entrepreneurship and startup education, Collaborative literacy education, maker's experience, adaptive learning system, and digital literacy education platform.

Keywords EduTch · Untact · Lifelong education · New normal 2.0

1 Introduction

With the advent of exponential growth due to the 4th industrial revolution, changes and innovations that have not been experienced throughout society such as extreme adjustment of industrial structure, mass unemployment, and polarization are becoming reality one by one. The new era demands new talents, and for this, a new educational paradigm is urgent. With the advent of a superintelligent, hyper-connected society, the traditional concept of school education is changing greatly. Ironically, it was COVID-19, not the CTO or CEO of mega IT companies such as Google, Apple, Microsoft, and Amazon, but the CTO or CEO of giant edutech companies, ironically, that triggered the digital transformation in the education field.

C.-J. Lee (✉) · S.-W. Choi
Department of Lifelong Education, Soongsil University, Seoul, Korea
e-mail: turby60@ssu.ac.kr

S.-W. Choi
e-mail: choiss@ssu.ac.kr

© The Author(s), under exclusive license to Springer Nature Switzerland AG 2021 191
H. Kim and R. Lee (eds.), *Software Engineering in IoT, Big Data, Cloud and Mobile Computing*, Studies in Computational Intelligence 930,
https://doi.org/10.1007/978-3-030-64773-5_16

With the spread of COVID-19 and the Untact Trend, the education industry is accelerating digital transformation with keywords such as self-directed learning capabilities, creativity, collaboration, and job skills.

Elvin Toffler said The illiterate of the twenty-first century will not be those who cannot read and write, but those who cannot learn, unlearn, and relearn [1]. In our society, where vast numbers of knowledge and information are exploding and the knowledge and information are evolving on their own, we must accept that it is not an age of education but an age of learning.

The edutech-based lifelong education system is the most pressing alternative for democratizing education and improving the quality of education by resolving polarization in education. In this paper, by looking at the trends in the AI and EduTech industries, and establishing a New Normal for lifelong education according to the spread of Untact Trends, the national lifelong education system can become the most basic means and methodology of education and learning that will move the future.

2 Explanation of Terms

New Normal New economic order and standards emerged after the 2008 World Financial Services Commission. A previously unfamiliar or atypical situation that has become standard, usual, or expected (Oxford Dictionary). It is also referred to as New Normal 2.0 because the changes that appear due to COVID-19 will become the new standard.

Digital Transformation Applying digital technology across society to innovate traditional social structures. In general, it means to innovate existing traditional operating methods and services by building and utilizing information and communication technologies (ICT) such as Internet of Things (IoT), cloud computing, artificial intelligence (AI), and big data solutions as a platform in enterprises (IT Glossary, Korea Information and Communication Technology Association).

Self-Directed Learning A form of learning in which the learner voluntarily selects and decides the entire process of education, from whether or not to participate in learning, to setting goals, selecting an educational program, and evaluating education.

Informal Education A general term for education that can occur outside of a structured curriculum [2]. It can refer to various forms of alternative education, such as unschooling or homeschooling, autodidacticism (self-teaching), and youth work. Informal education consists of an accidental and purposeful ways of collaborating on new information [3]. It can be discussion based and focuses on bridging the gaps between traditional classroom settings and life outside of the classroom [3].

Flipped Learning Contrary to the conventional teaching method, it refers to a class method in which students learn the lecture video provided by the professor in advance before class, and debate or solve assignments in the classroom. Flip-learning can be characterized by a flexible environment, a 'learning culture', 'intentional learning content', and 'professional educator'.

Micro Learning It means micro + learning, a short study that conveys an idea on a topic. Micro-learning is about engaging learning participants in a single action that is purposefully designed to drive specific outcomes. [Source] Edutech instructional design: micro-learning and adaptive learning concepts.

Adaptive Learning E-learning that provides learning information and methods according to the learner's level and learning style. It is to provide customized education by understanding how much students understand the class content and what they are good at. ICT technologies such as cloud services, big data, artificial intelligence, and virtual reality are used for adaptive learning.

3 The Fourth Industrial Revolution and Changes in Education

3.1 Fourth Industrial Revolution and the Spread of Untact Trends

The term Fourth Industrial Revolution was first mentioned by Klaus Schwab at the Davos Forum in January 2016, and is based on the digital revolution, ubiquitous mobile network, artificial intelligence and machine learning. As an industrial structure characterized by breadth and depth and systems impact, he explained that there will be a radical change in the social system, science and technology, and job structure.

The COVID-19 pandemic caused not only economic activity contraction, but also social issues such as education and learning gaps such as unprecedented online school opening and postponement of SAT. Education platforms previously designed and implemented in response to the need for an Untact that started with COVID-19 were rapidly and naturally put into public education and higher education sites. In addition, innovative and future-oriented educational models that have been theoretically suggested are spreading at a rate above the COVID-19 pandemic, entering the educational field, and converging with Edutech platforms to innovate school education. In the education industry, the Untact Trend created an opportunity for edutech to spread due to the voluntary and inevitable demands of teachers and learners, not led by IT companies or governments.

Teaching and learning methods in educational institutions are also constantly innovating in various forms. The spread of flipped learning, micro-learning, and adaptive learning, which are applied in various ways as an alternative to existing education, stems from the development and application of related Edutech technologies, and at the same time, leads the innovative growth of the Edutech industry. It is also doing.

3.2   Change in the Educational Environment

No one can deny that education is the area where the biggest change is expected to occur due to the 4th industrial revolution and advancement of AI. The educational paradigm in the era of the 4th industrial revolution requires fundamental changes in many areas. In the traditional way of teaching in the form of a school where teachers teach and students learn, cultivating talented individuals who can accept and utilize the enormous amount of diverse knowledge and information pouring out every day in the exponential growth era is physically. Since it is inevitably impossible, a new paradigm education is needed for a superintelligent hyperconnected society.

Education to acquire knowledge in areas where humans could not be better than artificial intelligence and machines may no longer be necessary. Acquiring knowledge and analyzing data is left to the outside world, and education aimed at leading society and solving problems creatively should be carried out by humans with wisdom that machines do not have and insight through vast knowledge. Wisdomization of life for the development of creative and emotional areas Education that cultivates insights penetrating knowledge and fosters cooperative spirit, communication, and empathy as a social being is education for the future, and education technology for fundamental change in educational paradigm Access should be attempted.

Changes in the economic structure and educational environment require new talents. Since people are ultimately responsible for R&D and application of new technologies, nurturing human resources is more important than anything else. First of all, it is necessary to cultivate subjects for research and development of science and technology related to the 4th industrial revolution and the flow of technology development is reflected in education reform, rather than a structure in which technological change changes the content and format of education. In addition to the ability to quickly acquire and develop knowledge, we must become creative talents who can understand, converge and develop new times and technologies. At the 2016 World Economic Forum, the 4th Industrial Revolution will have an impact on Skills Stability, and for this purpose, the ability to solve problems through communication and convergence, cognitive ability, and sensitivity and creativity with knowledge in IT technology and STEM fields. It was also predicted that talented individuals who can prepare for changes and actively adapt to change will be recognized as competent talents.

It is an era in which we must accept a variety of knowledge and information amid Exponential Growth, and live in a constantly updated knowledge and platform. It is self-evident that it is impossible to live a whole life only with public education and higher education that have to digest the curriculum set for a certain period of time. Even now, there is a limit to working life, entrepreneurship, and other economic activities only with the knowledge learned in the higher education curriculum. Companies are constantly investing in retraining for workers. In addition, individuals who are members of society who engage in any other economic activity, whether starting up business, must learn throughout their lives to understand new social, economic, and cultural paradigms and platforms in order to adapt to an ever-evolving society. This means that non-formal education along with formal education should be carried out throughout life, rather than formal education, the importance and weight of non-formal education will gradually expand. Therefore, it can be said that one of the new educational paradigms is that learners' will and literacy for non-formal education must be prepared through formal education.

4 Education Industry Trends and Education Innovation Cases

4.1 Education Industry Trends

The educational crisis triggered by COVID-19 accelerated the digital transformation with keywords such as self-directed learning capability, creativity, collaboration, and job skills. The core of the changes in the education industry following the Fourth Industrial Revolution can be said to be the innovation of consumer (learners)-centered education services, and the education industry is taking a new leap based on the following four trends. The first is a realization that stimulates the learners' five senses, and the second is a connection education that interacts anytime, anywhere using a platform based on super-connectivity that transcends time and space. As knowledge and information become increasingly decentralized, the spread of sharing economy service platforms that anyone can become both a consumer and a supplier of education and on-demand services that maximize educational effects by continuously and intensively grasping the needs of trainees are emerging. Third, it is an intelligent trend that conducts classes based on artificial intelligence and big data technology, and provides customized problems and feedback in consideration of subject achievement. This is changing the appearance of the classroom in a form in which a teacher implemented by artificial intelligence leads the class based on the curriculum, the teacher manages learning, manages achievement, and motivates learning, and in the field of the class leads discussion and collaboration. Finally, the convergence-oriented education platform business model is accelerated. The educational service platform centered on teachers, learners, and customer values based on 'intelligence' and 'connectivity' is promoting the expansion and growth of numerous businesses that will emerge not only from the education industry but also from the convergence between industries.

The Office of Science and Technology Policy (OSTP) at the White House announced 'Preparing for the Future of Artificial Intelligence' in October 2016, and researchers and researchers with interdisciplinary convergence knowledge A policy for fostering experts was prepared. Therefore, the National Academy of Sciences (NAP), the National Science and Technology Commission (NSTC), and The Committee on Science, Technology, Engineering, and Math Education (CoSTEM) are the science and technology intensive training programs for artificial intelligence education, data scientists. Through the development of a major curriculum for training, it has been striving to foster artificial intelligence manpower and to strengthen its capabilities and respond to the long-term increasing demand for future talent.

The French Ministry of Higher Education, Research and Innovation (MERSRI) has published a future strategy report containing about 50 policy recommendations to cope with the era of artificial intelligence. The ultimate goal of French education policy is to establish an AI ecosystem consisting of education-jobs-R&D-AI platform (community), and to maximize the competitiveness of the 4th industry. In particular, in the field of education, artificial intelligence education materials are provided at the national level in order to establish a curriculum related to artificial intelligence and data processing throughout the entire elementary and secondary education process, establish an artificial intelligence education ecosystem, and promote convergent research with other fields such as humanities, law, and sociology.

Korea defines education innovation as the most important educational policy, and in February 2018, proposes policies to innovate elementary and secondary students centered on school education, create college knowledge, and strengthen local growth capabilities to achieve the national goal of "cultivating a new future for Korea". Elementary and secondary education introduces the credit system and free grade system, strengthens software and STEAM education, and in the case of higher education, expands support for creative basic research and reorganizes the curriculum to strengthen job-hunting capabilities, and in the field of lifelong education, sustained growth in response to social changes.

According to the domestic edutech technology roadmap announced by the Small and Medium Business Technology Information Promotion Agency in 2017, domestic edutech technology is largely developing in three areas: realistic education, customized education, and software coding. Through open online classes, education at a global level is possible. In addition, platforms that enable customized learning with the development of AI technology, learning big data analysis, MOOC learning data analysis, learning agent, and social learning content technology are leading the innovation of learning conditions.

4.2 Examples of Innovation in School Education

Prior to the COVID-19 outbreak, the entire world had already predicted and prepared a hyper-connected society where everything was connected through the rapid development of advanced science and technology and artificial intelligence. However, due to COVID-19, innovative tests are faced with a reality in which generalization must be accelerated. Institutions that have innovated school education have common keywords such as competence, customization, self-directedness, communication and collaboration, problem solving, and creativity. The following innovation examples should be reviewed throughout our primary, secondary, tertiary and continuing education systems.

- **Alt School**—A lab school centered on personalized-learning models, digital literacy, and experiment-based learning. Support individualized curriculum for each student, and encourage teachers to motivate students to learn themselves. Practices typical STEM education, and is always evaluated through real-time feedback and interaction between teachers and students.
- **Khan Lab School**—Completion of personalized training recommended based on artificial intelligence technology. Classification and promotion based on autonomy and academic achievement. Classes are conducted in the form of a project that emphasizes collaboration. Learning through active communication with teachers, colleagues, and oneself. Learning based on context, such as cognitive ability and personality, not content-oriented. Autonomous learning through guidance from an advisor. Self-directed, small group, project-based routine, not daily progress by subject.
- **Summit School** [4]—Provide personalized learning. 1:1 mentoring with teachers, and group mentoring support. Operate by suggesting elements of content knowledge, habits of success, expeditions, and cognitive skills.
- **Minerva School**—Students are selected by evaluating cognitive skills and essays. The first-year common curriculum includes effective communication, critical thinking, the expression of imagination, and mutual exchange. The majors are divided into arts and humanities, computer science, natural science, social science, and business administration. There is no formal lecture class, and learning is conducted through flipped learning, and class participation is used as an important item for evaluation. Through an online virtual classroom platform called Active Learning Forum, subject achievement and learning participation are performed through an artificial intelligence system. Received on-site training for global internships through educational cooperation with global companies such as Google, Amazon, and Uber.
- **Fairfax County**—The educational innovation initiative is FCPSOn (Fairfax County Public School On), which is cultivating practical human resources suitable for the future society, literacy education as a digital citizen, providing equal access to learning, cultivating holistic talents with empathy, and professional educational tools for teachers. It presents a policy vision such as provision.
- With the introduction of a mobile academic achievement management system, parents and teachers can always check the student's academic achievement in real time, and the introduction of digital textbooks allows students to prepare and review anytime, anywhere.
- **Ecole 42**—Privately-led experimental global IT talent academy established with the aim of fostering startups. Classes are conducted in such a way that students solve technical tasks that occur in the actual corporate field based on team projects. Because they do not lecture, there are no professors, no textbooks, and no tuition fees. Although they do not have degrees, students score points through mutual evaluation, and most of Ecole42 graduates are employed by global IT companies. The educational goal is to develop professional abilities in creativity, cooperation, technicality, and sincerity, and is made through thorough team-oriented project work.

5 Edutech Industry Trends

5.1 Edutech Technology Trend

In the edutech industry, various types of educational platforms and services such as mobile and cloud services are being developed according to the characteristics of consumers. In addition, learning services that incorporate interactive production technology based on AR and VR-based educational contents are increasing. It is leading the improvement of the education system with a national initiative in order to cultivate key talents for intelligence such as SW, AI, big data, and cyber security, and to cultivate convergent human resources to produce realistic content such as AR and VR. To this end, the development and diffusion of STEAM educational models for elementary and secondary schools, the establishment of an intelligent teaching and learning platform, the establishment of an open market for educational contents, and expansion of K-MOOC education will be promoted. It focuses on dissemination, expansion of wireless infrastructure, and strengthening of teachers' SW capabilities.

The technologies that should be noted most as edutech related technologies are AR/VR, AI, robotics, and blockchain. AR/VR is expected to grow 7 times from $1.8 billion in 2018 to $12.6 billion in 2025. AI, which will implement customized education, teaching methods, and administrative management automation services through big data, is also expected to form a market of $6.1 billion in 2025 from 800 million in 2018. VR/AR-based learning technology can maximize learning effects through audio-visual, olfactory, hearing, and tactile senses such as virtual reality, augmented reality, hologram, and five sense media, and related hardware technologies include HMD (Head Mounted Display) and interaction HW. It is implemented with 360° photographing technology of a place or object, computer graphic (CG) technology, etc., photographing equipment, measuring equipment, etc. that create virtual space.

Edutech related technologies can be divided into implementation technology and provision technology as shown in the following table.

Division		Contents
Implementation technology	Web-based content	Web-based educational contents such as e-learning and lecture videos
	Mobile learning content	Educational content provided based on smart devices such as smartphones, tablet PCs, and wearable devices
	Social learning content	Educational contents provided based on SNS such as Facebook, Kakao Talk, and Line
	Simulation-based virtual reality content	Content provided to experience sensations such as cigars and hearing in a virtual reality environment
	Augmented reality contents based on spatial recognition	Educational content that can be experienced beyond time and space as a location-based service
	Hologram based on realistic image	Learning contents created using the principle of holography
	Five senses interaction content	Learning contents expressed in electronic form by combining codes, text, audio, images, images, and five sense information with ICT technology
Provide technology	Native computer programming	Learning contents produced in the form of simple games or animations by connecting scripts as if assembling blocks
	Personalized Edutech content	Adaptive learning that provides educational content tailored to personal scores and learning tendencies through mining, analysis, machine learning, and deep learning of learners' data
	Learning agent	Data analysis and mining technology that analyzes learners' learning behavior patterns and provides learning courses, methods, and strategies suitable for learners
	MOOC learning data analysis	Technology that analyzes and evaluates learning patterns and knowledge perspectives experienced by learners on large-scale contents, learning materials, and interactions provided by MOOC

[Table 2] Edutech related technologies

5.2 Edutech Market Trend

As the influence of mobile devices and content increases, the learning app and media market is growing rapidly. From existing e-learning companies to start-ups with various technologies (early stage companies), the market is growing rapidly by jumping into Edutech. STATISTA predicts that the global learning software and app market will grow rapidly from 2017, with an annual growth rate of 2.2% until 2022 (Citation, Institute for Information and Communication Policy, Growth of the Learning App Media Market 2019). Research and Market predicted that the U.S. educational app market will grow at an annual average growth rate of 28.15% from 2016 to 2020. Not only that, but also a new new on YouTube There is also an increasing number of cases where knowledge or skills are learned or used for regular learning purposes.

Learning apps and media are expected to create new forms of added value and grow by converging with other industries and fourth industry related technologies such as wearable technology and VR/AR technology. In addition, it is expected to be able to solve social problems such as the gap between the rich and the poor in education and private education, as well as the growth of the ICT industry.

As countries around the world are reinforcing mid- to long-term education policies and technology development from infants to adults, there is a growing demand for technology development related to customized creative education technology and interactive interactive education service. The United States is planning to digitize all classrooms and is preparing to finance an educational software standard called Tin Can Project, which enhances personal learning management lifelong education. The UK has established Edutech UK, and Southeast Asian countries are also rapidly shifting their e-learning policies to introducing Edutech and making efforts to build content and platforms. Universities at home and abroad are also striving to improve the quality of MOOCs, and are leading the expansion of high-quality online content at the national level.

The Edutech industry is developing into various types of educational platforms and services that take into account the characteristics of consumers based on technologies such as mobile and cloud, open source learning platforms, and national online content distribution. In addition, in the public education and private education markets, it can be said that the demand is great not only in elementary, middle and high school online learning, specialized lectures above college, but also in individual education in the fields of language and hobbies, and job education for companies and organizations.

The US edutech market is dominated by Google, Apple, and Microsoft. Google is expanding the software market as well as hardware Chromebooks. It also introduced Google Classroom, a cloud-based education platform, and G Suite for Education, Google Expedition. Apple is targeting the education market through the iPad and Apple Classroom, and Microsoft is actively engaged in M&As related to education, the video platform Flipgrid, the task sharing and topic discussion service, Chokeup (CHalkup) is being provided to provide services, and is striving to preoccupy the education market with Office 365 Education. Amazon is also servicing "Amazon Inspire," a platform for sharing learning materials necessary for elementary and middle school students.

Although the domestic edutech market is expanding, the growth trend is low, and large businesses with sales of 10 billion won or more account for only 3% of the total number of business operators. It is characterized by existing education service companies cooperating with competitive Edutech startups or introducing Edutech into their services. Daekyo acquired the AI math education platform to provide learning services, and Woongjin Thinkbig acquired a stake in a machine learning startup to provide a service that analyzes learners' behavior patterns using big data and AI technology. The faculty developed a computer coding education service in partnership with a coding education robot manufacturing startup.

The adoption rate of AI in education is still low compared to other fields. Artificial intelligence, which enables the automation and intelligentization of customized education, teaching methods and administrative management through big data, is also expected to form a market of $6.1 billion in 2025 from 800 million in 2018. [Source] 2020 Edutech Industry Market Size, Trends and Implications|Author Reallog.

6 New Normal of Lifelong Education According to the Spread of Untact Trend

Along with the dailyization of technologies related to the 4th industrial revolution, COVID-19, Untact Trends, and distancing became the most important reasons to completely re-establish educational goals and systems. The key keyword in the era of New Normal 2.0 is "speed and adaptability." Non-face-to-face society and deglobalization. It is more important than anything else to define an image of talent that meets the needs of the rapidly changing era, and to establish a New Normal and innovation task for lifelong education. Creativity (creativity), critical thinking (critical thinking), challenge (challenging spirit), convergence (convergence capability), collaboration (empathy and collaboration), curiosity, etc. Accordingly, the educational system should also be accompanied by innovative changes. Learned creativity, critical thinking, and convergence capabilities will no longer work, and insights based on learning and understanding have become more important than knowledge.

Already we are living in the era of lifelong education and lifelong learning. You have to constantly learn, and if you don't constantly innovate, you will be cut off. Motivation in everyday life, not through education, should be the aim of lifelong education. Learning is an act that can be done without a teacher. Edutech has become a learning opportunity in the crisis of education for lifelong learning. Ultimately, it is necessary to re-establish the life cycle and educational goals of the existing lifelong learning, which focused on adult learning, in line with the new era. For this, I would like to present some New Normals from the perspective of lifelong education.

- Lifelong education should be redefined as a classification system centered on future social demands and talents, and the purpose of lifelong education should be the realization of literacy education, which is the foundation for innovative growth. In the current Lifelong Education Act, Article 2, No. 1, lifelong education includes education supplementary education, adult character acquisition education, vocational competency improvement education, humanities education, culture and arts education, and civic participation education, excluding the regular school curriculum. It is said that it refers to all types of organized educational activities. The knowledge and literacy of the regular curriculum, which has a single subject mainly aimed at acquiring knowledge, curriculum-centered education, and evaluation system, has already been established in which an environment in which self-directed learning can be learned through the process. This is because it is an era in which people can directly access the education of the world's best scholars and experts from short stories through online. Rather, Creativity, Critical thinking, Challenge, Convergence, Collaboration, and digital literacy are defined as a lifelong education classification system. The reorganization of a system for learning educational contents by level should be discussed.
- It should be a social infrastructure for self-directed learning ability. In a super-intelligent, hyper-connected society, it is difficult to continue economic activities without constant learning and learning. Continuing education has become the only alternative to overcoming the school education frame that had dominated for

decades. We must accept that the age of technology and knowledge overflowing and evolving itself is not an age of education, but of lifelong learning. Until now, education has been centered on the instructor, that is, the supplier. However, the people the future society demands are those who learn and learn. Through the so-called 'Teach less, Learn more' educational method, you can become a talented person and self-development person required by society.

- You shouldn't simply educate yourself on self-directed learning skills. You need to cultivate an active learning literacy that allows you to find your own needs and needs for learning and to motivate yourself. Curiosity and self-directed learning skills are the most necessary for sustained growth and self-development.
- Future-oriented entrepreneurship and entrepreneurship education should be the core theme of lifelong education. Creation and innovation have become one of the most important virtues for humanity in the future. The future-oriented entrepreneurial spirit is not just about the attitude or spirit that an entrepreneur should have to pursue profits and fulfill social responsibility. Mind to continuously innovate, organizing, implementing, and risk-taking innovation to realize business opportunities, organizational and time management skills, patience, rich creativity, morality, goal setting ability, appropriate sense of adventure, sense of humor The core theme of lifelong education should be the ability to deal with information, the ability to conceive alternatives for problem solving, creativity to generate new ideas, decision-making skills, and entrepreneurial literacy. Through this, know-what education to understand what is the problem and the ability to create more jobs by themselves should be strengthened. The literacy education to foster field-oriented human resources that fits the newly changing society, economy, and market order should be the standard of lifelong education.
- Lifelong education should contribute to transforming a society where all members are equipped with the knowledge of collaboration. Artificial Intelligence, Robot, Big Data, Drone, Autonomous Driving, Virtual Reality, Augmented Reality, Convergence Reality, Internet of Things, Brain-Computer Interface, Fintech, Food Tech, Quantum Computer, 3D Printer, Blockchain, Biotechnology, Android, Cloud computing, edge computing, smart mobility, nuclear fusion, etc. No individual can handle all of these broad areas of knowledge and experience. Even if a problem has been found and an idea to solve social and system issues through problem definition, it is impossible to solve it "alone". This is why collaboration between members is necessary, and smooth communication is difficult even if only certain individuals have awareness of collaboration. Humans are social animals. Throughout my life, I live in constant relationships with others, and that will not change in the future. Being an imperfect human, he needs the help of other members of society. The more you go to a higher school, the more subdivided and narrower the area of study is. In the coming New Normal era, it will be subdivided into more diverse majors, social and economic structures, and the more advanced collaboration skills are needed. In order to solve more diverse and complex problems in the future, in order to solve everyday problems and pursue future values, the knowledge of collaboration is essential and must be continuously nurtured in the category of lifelong education.

- Lifelong education should provide a place to gain makers experience. The era of making a lot and consuming a lot is over. In a society where culture, society, economy, and consumption structures are gradually personalized and Untact is preferred, large-scale production and consumption systems of large corporations are being replaced by artificial intelligence, machines, and online systems. Instead, more and more consumers are looking for something that's valuable to them and their own. Craftsmen who thought they only disappeared in the era of the 4th industrial revolution and artificial intelligence, 'gold hands' from all over the world are appearing. Not only are the products made by the craftsmen's "thinking hands" of high quality. There are many products that have solved the inconvenience of existing products.
- In order to maintain and develop this changed future market economy structure, public literacy must be generalized. Experience with Makers is the social and economic value of the future and must be cultivated through lifelong education. From first-hand ideas to design experience, aspiration for items, conscious training and honedness, and a spirit of challenge for new things will be equipped through Makers' experience.
- An adaptive learning system through adaptive learning and micro-learning should be applied to the lifelong education field. In traditional curriculum learning, everyone has performed the same amount of learning in the same order. Since it was important to convey the same content of knowledge to any target, all education was conducted and evaluated regardless of the learner's level, disposition, and learning attitude.
- With the advent of the 4th industrial revolution era, digital transformation took place, and customized learning became important according to each individual's competence and level. Micro-learning is being emphasized more as a small unit of learning suitable for individuals is required rather than the same knowledge and progress. Personalized learning is possible only when small learning contents exist. The development of Edutech enables adaptive learning, which is personalized learning, and has created an environment in which learners can freely choose the content and learning method they need.
- New educational experiments are in full swing not only in elementary, middle and high schools around the world, but also in higher education fields. In the continuing education market, individual differences, dispositions, characteristics, and requirements for learning levels are much more diverse. There is no longer a need for uniform education by collectively gathering learners of different levels of similar individual curriculum in numerous lifelong education institutions. It is necessary to establish a system that can continuously produce high-quality personalized learning topics, curriculum, and contents in the field of lifelong education.
- Lifelong education should form a platform for digital literacy education [5, 6]. Digital literacy refers to the ability to use digital as well as to read and write text. The speed of development and spread of IT and AI technologies will produce enormous digital illiteracy. The way to minimize digital illiteracy and grow into a citizen who will be the target of new production activities is to continuously

apply a digital literacy education platform. Lifelong education should contribute to equipping a democratic future civil society by establishing a digital literacy learning system throughout the life cycle.

Digital literacy education does not mean functional and technical education such as information education in the past. In the past, it was defined only as the level of ability to understand information in the sense of consultation, but in recent years, it includes critical thinking, problem-solving ability, creative expression of one' thoughts and knowledge, and the ability to create content. In addition, it is defined as digital citizenship, communication, and collaboration beyond the functional capabilities of individuals. The Digital Literacy Education Association divides the digital literacy education framework into separate areas, in addition to digital ethic education and digital ability education, and digital application education. Unlike other literacy, digital literacy regards social participation as important, and social participation is increasing through the development of digital media and technology, so whether it is practically applied beyond having ethics and capabilities was considered important. Personal and social attitudes such as appropriateness of digital use, self-development use, self-management such as self-control and self-control, sharing information and knowledge, public interest, and altruistic use of digital were emphasized. In digital communication literacy, relationship management, crisis management, It is noteworthy that conflict management was added.

7 Conclusion and Future Research

There are many discussions about the trends and future prospects of lifelong education, but it seems that the existing purpose of lifelong education and fundamental topics such as classification system have not been sufficiently reviewed. In this paper, we investigated the trends of artificial intelligence and edutech in the education market, and presented a New Normal from the perspective of lifelong education according to the spread of Untact Trends. Although it has been only four years since the discussion on the 4th industrial revolution began, the world was already on the verge of exponential growth, hyperconnected society, and entry into a superintelligent society through the rapid development of advanced science and technology and artificial intelligence. The COVID-19 pandemic accelerated social change across politics, society, economy, culture, education, and art society, and the spread of Untact Trends caused the necessity of re-establishing New Normals across society.

Edutech is fully introduced and spread to the fields of elementary, middle and high school formal education and higher education, the most conservative industries. Innovative teaching and learning methods around the world are being applied in the field of education, breaking down the existing framework of education. Artificial intelligence technology has already been developed to a level that can implement and apply the educational system desired by future education. Education infrastructure such as edutech learning platform and contents applied with artificial intelligence will become better teachers, mentors, and facilitators than any instructor.

The purpose and concept of lifelong education, as well as the educational system, must be changed to a paradigm suitable for a superintelligent and hyperconnected society. Lifelong education transforms the educational purpose of 'continue the individual's physical and personal maturity and social, economic and cultural growth and development throughout life' into 'educational infrastructure to foster democratic citizens who lead the future society' shall. To this end, this paper presents a New Normal of lifelong education as a lifelong education classification system innovation, self-directed learning ability cultivation, future-oriented entrepreneurship and startup education, Collaborative literacy education, maker's experience, adaptive learning system, and digital literacy education platform.

In the future, we will study the role and weight of the digital literacy education framework and theoretical and technical measures for application to lifelong education platforms.

References

1. Future Shock. Mass Market Paperback. by Alvin Toffler, Bantam Books, 07 2013
2. Rogoff B, Callanan M, Gutiérrez K, Erickson F (2016) The organization of informal learning. Rev Res Educ 40:356–401
3. boyd, d (2013). It's Complicated the Social Networking Lives of Teens. https://yalebooks.yale.edu/book/9780300199000/its-complicated Archived 2019.7.2.
4. "'A Bit Of A Montessori 2.0': Khan Academy Opens A Lab School". NPR.org. Retrieved 07 Aug 2016
5. Khan Lab School reinvents American classroom, retrieved 28 July 2017
6. "Minerva Project and KGI Partner to Launch the Minerva Schools at KGI" (PDF). Retrieved 24 Jan 2015

The Optimal Use of Public Cloud Service Provider When Transforming Microservice Architecture

Sungwoong Seo, Myung Hwa Kim, Hyun Young Kwak, and Gwang Yong Gim

Abstract The microservice architecture (MSA) is an architecture that divides a single application into a number of small applications to facilitate change and combination. When application deployment takes a long time, partial function modification affects the whole, or overall Q/A is requested, architecture can be configured by dividing into microservices for each function. As described above, microservices are small units that are aimed at individual application computing in an operating system (OS) environment, not computing in a virtualized environment. And because it is not related to the virtualization environment (type of hypervisor), it is not affected by the virtualization environment of the user's data center (IDC) or the type of public cloud service provider (CSP). In addition, it has the advantage of being able to create, modify, and delete to the cloud by using a method of distributing users' on-premise microservices to the outside. It can provide an optimal environment even in situations where many reviews of hybrid cloud for backup, recovery, and expansion purposes of enterprises (utilizing infrastructure transition between on-premises and public cloud). In this paper, we analyzed the functions of CSPs in terms of functionality when transferring microservices to an external public cloud in the context of configuring such microservices in the user environment (on-premise, linux). As a result, the features and usability of CSPs were verified, and the application plan according to the specific business environment of companies was explained. In addition, the contributions of this thesis for future research are explained.

S. Seo · M. H. Kim · H. Y. Kwak
Department of IT Policy and Management, Soongsil University, Seoul, South Korea
e-mail: cpim.seo@hanwha.com

M. H. Kim
e-mail: beauhwa1@naver.com

H. Y. Kwak
e-mail: khykhan@naver.com

G. Y. Gim (✉)
Department of Business Administration, Soongsil University, Seoul, South Korea
e-mail: gygim@ssu.ac.kr

© The Author(s), under exclusive license to Springer Nature Switzerland AG 2021
H. Kim and R. Lee (eds.), *Software Engineering in IoT, Big Data, Cloud and Mobile Computing*, Studies in Computational Intelligence 930,
https://doi.org/10.1007/978-3-030-64773-5_17

Keywords Cloud computing · Container virtualization · CSP (cloud service provider) · MSA (microservice architecture) · Docker · Kubernetes · DevOps

1 Introduction

Currently, the system resource environment in most data centers (IDC) uses a hypervisor-based virtualization cluster as an architecture and is maintained through a series of operations that create, delete, and modify VM (Virtual Machine). The environment evolves from a traditional infrastructure architecture (router-switch-server-storage) called a legacy configuration to a level of virtualization of the server level. It was settled in a way called 'server virtualization', and technically, it brought great advantages in terms of saving networking interfaces while improving the space and power consumption of hardware. However, the waste of remaining resources peculiar to full-virtualization/para-virtualization and simultaneous resource exhaustion that occurs when many requests are made are a chronic problem that is difficult to solve with the server virtualization method. It has weaknesses such as a static structure that is not free from scale-out (a method of expanding a system by adding multiple servers). This weakness has led to the illusion that, in the short term, the public cloud may become a substitute for on-premise systems in the infrastructure market [1]. At the same time, it was an opportunity to bring about competition in the server virtualization market, which was the next generation, but became conservative like the legacy.

The aforementioned weaknesses of server virtualization, such as measures to increase the reuse rate of the remaining resources within the on-premises system, free scale-out, and automation, were being reviewed. From these concerns, the technology of a container that allows applications within the server to be used like an OS, rather than server virtualization, emerged in a competitive situation [2].

From the legacy of the above to the current virtualization and public cloud, the key points that can be seen from reviewing the evolution of infrastructure can be summarized in a few. The first is development by demand. The server virtualization system emerged from the legacy configuration, where all resources had to be physically configured, and improved for efficiency such as physical space and power. Even after the server has been virtualized, the advancement to a more microscopic direction is in progress with SDN (Software Defined Network) and containers due to the demand for efficient use of system resources [3]. The second is the service from the user's point of view. Users should receive the same service no matter what system the service they use operates on, and they want to continue to operate in that experience. The third is stability. Service has to respond to unplanned system outage due to a failure or a service affecting the system by many requests from users. In addition, it should be possible to ensure stable service resumption for measures to quickly take measures such as improvements due to bugs in service applications [4].

It is clear that public cloud services are at the peak of the technology in the light of the development process of the above infrastructure [5]. However, due to various concerns such as security, platform lock-in, and cost, many companies suddenly discard internal virtualization resources and cannot migrate to the public cloud [6, 7]. Therefore, we have no choice but to think about ways to efficiently mix and use services that operate in the legacy, services that operate in virtualization, and numerous applications with the public cloud. The container-based microservice architecture (MSA) at this time can be suggested as an optimal alternative to this [8, 9]. It is a method of testing in the public cloud by prioritizing those that can be microservices with containers. For reference, in recent years, a management system called "Kubernetes" is in the spotlight as an open source that provides automatic distribution and scaling of containerized applications [10, 11].

Accordingly, in this paper, it is assumed that microservices are operated on a container basis. We analyzed the suitability of public CSP (Cloud Service Provider) to transfer common and popular Linux applications (e.g. apache, tomcat) from the user's on-premises to an external public cloud. In addition, microservices (applications) of both on-premises and public cloud were analyzed for functional aspects of synchronization according to user intention (Each CSP has a different name, but similar functions were analyzed, and detailed options of each function were not compared with each other). There are various global CSPs, and each vendor provides complex services with unique features. In this situation, the user's point of view seeks to ensure optimal performance (cost, time, stability, reliability, usability) that satisfies various considerations, but it is difficult to confirm the correct answer in advance. This study will help to optimal use of CSP when transitioning MSA (microservice architecture) from on-premises infrastructure environment to public-cloud service.

2   Literature Review

The infrastructure environment, architecture and virtualization technology related to this study were classified and reviewed.

A. *Infrastructure environment*

The On-Premise infrastructure environment is built using software directly on a server, and most systems have been configured in this way from the past. In recent years, as cloud computing technology has been developed, private/public cloud services are being used in connection with business environments [12].

Private Cloud A private cloud is a place where data is stored as a closed cloud and is stored on an internal server of a company that uses the service.

Public Cloud The public cloud is a place where data is stored as an open cloud and is stored on an external server of a company that uses the service.

The public cloud uses the service through CSP, and the world's leading operators are Amazon, Microsoft, and Google. There is a need for research on classification/analysis of service characteristics for various CSPs [13].

Fig. 1 Monolithic system [16]

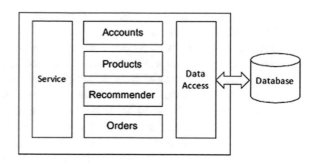

B. *Architecture*

In utilizing cloud computing, the use of virtualization technology in open source projects is emerging, which is a microservice-based structure, and as an example of a representative technology, Kubernetes (Management system that automatically deploys and scales containerized applications based on open source) has been applied a lot [14, 15].

Monolithic Architecture The monolithic architecture is an architecture structure that is generally composed from the past, and the developed environment is similar, so the complexity is low. However, if the scale of the project is expanded too much, the application run time increases and the build and deploy time increases [16] (Fig. 1).

Microservice Architecture Microservices develop service units for each function. Due to this, various advantages exist, which can be developed in different programming languages, can be extended independently from other services, and can be distributed on the hardware that best suits the needs. In addition, even if there is a single service failure, the entire system is not interrupted, so maintenance is easy and there is a fault-tolerant [16] (Fig. 2).

Compared to the monolithic architecture and microservice architecture styles, both have their pros and cons. In conclusion to Table 1, we can see that the style

Fig. 2 Microservice-based system [16]

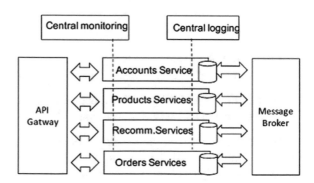

Table 1 Comparing monolith and microservices [17]

Category	Monolith	Microservices
Time to market	Fast in the beginning, slower later as codebase grows	Slower in the beginning because of the technical challenges that microservices have. Faster later
Refactoring	Hard to do, as changes can affect multiple places	Easier and safe because changes are contained inside the microservice
Deployment	The whole monolith has to be deployed always	Can be deployed in small parts, only one service at a time
Scaling	Scaling means deploying the whole monolith	Scaling can be done per service
DevOps skills	Doesn't require much as the number of technologies is limited	Multiple different technologies a lot of DevOps skills required
Understandability	Hard to understand as complexity is high. A lot of moving parts	Easy to understand as codebase is strictly modular and services use SRP(single responsibility principle)
Performance	No communicational overhead. Technology stack might not support performance	Communication adds overhead. Possible performance gains because of technology choices

of microservice architecture turns attractive when working with large codebases. For small projects, there may not be enough time to compensate due to the technical challenges that microservices bring. Also, if your team lacks DevOps (which means combining development and operation, which implies the concept of agile development and continuous integration), then stick with monoliths at first [17].

C. *Virtualization*

Hypervisor Virtualization Hypervisor-type virtualization is a type of virtualization engine and is software that allows multiple OSs to run simultaneously. Hypervisor, a software that realizes server virtualization, is used to support the hardware virtualization function [18] (Fig. 3).

Container Virtualization Container-type virtualization provides an independent OS environment by creating containers on the host OS running hardware. Library is located in the container, so it has the characteristic of being used as a separate server [18] (Fig. 4).

Container-based virtualization has attracted attention as a promising alternative, with the recent popularity of Docker (an engine that automates a container application), one of container virtualization technologies [19, 20].

Fig. 3 Hypervisor-based
virtual service [18]

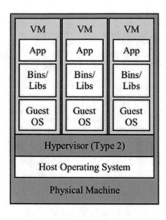

Fig. 4 Container-based
virtual service [18]

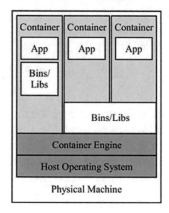

3 Comparative Analysis of CSP Characteristics When Transforming MSA

The status of major public cloud providers (CSPs) by country is different, and there are various global vendors (Table 2).

Table 2 Evaluated vendors and product information [20]

Vendor	Product evaluated
Alibaba	Alibaba cloud
Amazon web services	AWS cloud
Google	Google cloud platform
IBM	IBM cloud
Microsoft	Microsoft azure
Oracle	Oracle developer cloud

Fig. 5 Public cloud development and infrastructure platforms, North America, Q1 2020 [21]

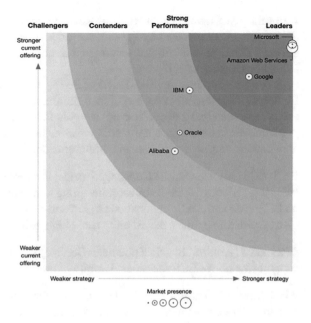

This study compares and analyzes detailed characteristics of Amazon's AWS (Amazon Web Service), Microsoft's Azure, and Google's GCP (Google Cloud Platform), the representative cloud services of major public cloud providers (CSPs) (Fig. 5).

A. *MSA-based infrastructure configuration*

To make the transition from on-premise and infrastructure to the public cloud quickly and simply, the initial infrastructure must be built with MSA rather than monolithic methods. This is why technology acceptance is necessary for MSA, which has recently been highlighted as a new technology, and the concept should be considered integrated with the application (S/W) and infrastructure (H/W). However, not all systems will reflect MSA and their application should be analyzed and classified according to the business environment.

B. *AWS (CSP1)*

AWS is the cloud service provider with the largest customer base. The microservices available items provided by AWS are as follows.

EC2 It is the most basically provided VM server on AWS, and you can use it by setting container-based microservices in EC2 (Amazon Elastic Compute Cloud). As it is a general purpose VM, it should be viewed as an infrastructure-as-a-service (IaaS) category, and when switching containers on-premise to public, if settings similar to the on-premise side in advance are required for the public cloud, they are almost within AWS. It is classified as the only customizable option. It is provided as a general computing resource. (You must select Linux system for microservice).

Lambda AWS Lambda is an event-driven (programming paradigm in which the flow of programs is determined by events), serverless computing (Serverless computing allows a cloud provider to run servers and dynamically manage machine resource allocation) platform [22].

It is a function-as-a-service (FaaS) of AWS, and it provides a container service by AWS. Therefore, the user can upload the function of the container to AWS and use it immediately. However, it is not suitable for the transition from on-premise to public cloud because there are restrictions on the use of special DBs, customized applications, etc., and the operating width of users is limited.

ECS ECS (Amazon Elastic Container Service) is a container package (provided with EC2) that reduces the hassle of users installing and deploying container applications on EC2 VMs and can be understood as platform-as-a-service (PaaS). Since there are many pre-define options, the usability is not high in light of the purpose of this study.

EKS EKS (Amazon Elastic Kubernetes Service) provides a Kubernetes cluster that supports the above ECS functions. Usability is the same as ECS.

AWS Characterization Results For the transfer (transition) of various CSPs for objective comparison research purposes, the most ideal model is when microservices (e.g. Docker and Kubernetes) are mounted on the basic infrastructure and then on-premise and public cloud are clustered. In this respect, it functions like on-premise. It is appropriate to use the EC2 (basic, Ubuntu) VM with the least restrictions.

However, in order to use Kubernetes quickly and easily, it is desirable to use basic PaaS using 'EKS' provided by AWS, and it has been confirmed that resources and packages can be configured in practice.

In addition, it may be better to use the services provided by default depending on the specific environment, constraints, and diversity/complexity of the used enterprise. EC2 provides a truly virtual computing environment, so you can launch instances on a variety of operating systems using a web service interface, load instances into a custom application environment, manage access to your network, and run your image using as many or few systems as you desire [23].

C. *Azure (CSP2)*

MS (Microsoft) Azure is a cloud service operated by MS and a cloud service provider with a large market share. The microservice available items provided by Azure are as follows.

Virtual Machine It is the most basically provided VM server in Azure, and it can be used by setting container-based microservices in each virtual machine.

Since it is a general purpose VM, it should be viewed as an IaaS category, and when a container on-premise is converted to public, it is classified as the only customizable option within Azure when a setting similar to the on-premise side is required in the public cloud in advance. It is provided as a general computing resource. (Microsoft's public cloud for microservices, but it must choose a Linux series).

Azure Functions It is a FaaS of Azure and container service is provided by Azure. Therefore, the user can upload the function of the container to Azure and use it immediately. However, it is not suitable for the transition from on-premise to public cloud because there are restrictions on the use of special DBs, customized applications, etc., and the operating width of users is limited.

AKS (Azure Kubernetes Service) It is similar to EKS of AWS, and it automatically provides IDC resources including virtual machines.

Azure Characteristics Analysis Results For the transfer (transition) of various CSPs for objective comparison research purposes, the most ideal model is when microservices (e.g. Docker and Kubernetes) are mounted on the basic infrastructure and then on-premise and public cloud are clustered. From this point of view, it is appropriate to use an Azure virtual machine (default, ubuntu) with the least functional limitations and restrictions such as on-premise.

However, in order to use Kubernetes quickly and easily, it is desirable to use basic PaaS using 'AKS' provided by Azure, and it has been confirmed that resources and packages are practically configurable.

In addition, it may be good to use the services provided by default according to the specific environment, constraints, and diversity/complexity of the used enterprise. Azure VM (Virtual Machine) is one of the scalable on-demand computing resource types provided by Azure. In general, you choose a VM when you need more control over your computing environment than the other choices offer [24].

D. *GCP (CSP3)*

Google Cloud Platform (GCP) is a cloud service operated by Google. The following items are available for microservice provided by Google.

Compute Engine It is the most basic form of general computing resource provided by GCP. As it is a general purpose VM, it should be viewed as an IaaS category. When converting a container in on-premise to public, if a setting similar to the on-premise side is required in the public cloud in advance, GCP can only be customized through this option.

GKE Google Kubernetes Engine (GKE) provides a management environment for deploying, managing, and extending containerized applications using Google infrastructure. A GKE environment consists of several machines (especially Compute Engine instances) grouped together to form a cluster [25].

Google Cloud Functions It is FaaS of Google Cloud and GCP provides container service. Therefore, users can upload containers to GCP and use them immediately. However, it is not suitable for the transition from on-premise to public cloud because there are restrictions on the use of special DBs, customized applications, etc., and the operating width of users is limited.

GCP Characteristic Analysis Results For the transfer (transition) of various CSPs for objective comparative research purposes, the most ideal model is when microservices (e.g. Docker and Kubernetes) are mounted on the basic infrastructure, and then on-premise and public cloud are clustered. From this point of view, it is appropriate to use a GCP Compute Engine (default, ubuntu) VM that has the least functional limitations and restrictions, such as on-premise.

However, in order to use Kubernetes quickly and easily, it is desirable to use basic PaaS using 'GKE' provided by GCP, and it has been confirmed that resources and packages can be configured in practice.

In addition, it may be good to use the services provided by default according to the specific environment, constraints, and diversity/complexity of the used enterprise. Compute Engine offers predefined virtual machine configurations for a variety of needs, from small general purpose instances to large memory-optimized instances or compute-optimized fast instances with lots of vCPUs (virtual Central Processing Units) [26].

4 Conclusion

This paper analyzed the adequacy of CSPs to transition common and popular Linux applications (e.g. apache, tomcat) from the user's on-premises to the external public cloud. In addition, we analyzed the effect of microservices (applications) in both on-premises and public cloud on the functional aspects of synchronization according to user intention. Microservices architecture can be viewed as a service that separates large projects into small, independent objects in terms of development. In addition, it can be interpreted that in terms of infrastructure, large applications from a monolithic perspective used to operate on high-capacity server systems, which translates micro-viewed individual modules into environments that operate on multiple server systems with low capacity. Still, it is the CPU, memory, and network elements that support usability for each object, but the computing method using them must be constantly considered. Therefore, this paper attempted to derive the most functionally appropriate CSP operators when transitioning from on-premise to public cloud by using container virtualization, which is the best 'infrastructure' at the present time for companies and users to utilize microservices. We derived the function of providing microservice architecture through containers for each CSP. In addition, it has been confirmed that in practical use of Kubernetes new technology to convert resources (moving between on-premises and public cloud), all can be utilized through specific services provided by major CSPs (limited to container usage conditions). It is possible to select a CSP according to a specific situation in each enterprise-level environment. It is judged to be a very meaningful result to compare and derive the provided functions from an integrated perspective and directly test and classify the usability for a number of major CSPs that are objectively recognized in the global market.

Based on this study, in future studies, it will be possible to specify performance indicators through tests for rapid service conversion and usability maintenance when on-premises containers are converted to public cloud under various environment configurations (including constraints). In addition, it will be meaningful to study the characteristics of each CSP utilization through the same environment configuration (e.g. VM-VM or container-container, etc.) for on-premises and public cloud.

Reference

1. Jatoth C et al (2019) SELCLOUD: a hybrid multi-criteria decision-making model for selection of cloud services. Soft Comput 23(13):4701–4715
2. Pérez A et al (2018) Serverless computing for container-based architectures. Future Gener Comput Syst 83:50–59
3. Gholipour N, Arianyan E, Buyya R (2020) A novel energy-aware resource management technique using joint VM and container consolidation approach for green computing in cloud data centers. Simul Modelling Pract Theor:102127
4. Taibi D, Lenarduzzi V, Pahl C (2017) Processes, motivations, and issues for migrating to microservices architectures: an empirical investigation. IEEE Cloud Comput 4(5):22–32
5. Chang DW, Pathak RM (2018) Performance-based public cloud selection for a hybrid cloud environment. U.S. Patent No. 10067780
6. Linthicum DS (2016) Moving to autonomous and self-migrating containers for cloud applications. IEEE Cloud Comput 3(6):6–9
7. Janarthanan K et al (2018) Policies based container migration using cross-cloud management platform. In: 2018 IEEE international conference on information and automation for sustainability (ICIAfS). IEEE, pp 1–6
8. Singh V, Peddoju SK (2017) Container-based microservice architecture for cloud applications. In: 2017 international conference on computing, communication and automation (ICCCA). IEEE, pp 847–852
9. Fazio M et al (2016) Open issues in scheduling microservices in the cloud. IEEE Cloud Comput 3(5):81–88
10. Linthicum DS (2016) Practical use of microservices in moving workloads to the cloud. IEEE Cloud Comput 3(5):6–9
11. Taherizadeh S, Grobelnik M (2020) Key influencing factors of the Kubernetes autoscaler for computing-intensive microservice-native cloud-based applications. Adv Eng Softw 140:102734
12. Mansouri Y, Prokhorenko V, Babar MA (2020) An automated implementation of hybrid cloud for performance evaluation of distributed databases. J Netw Comput Appl 102740
13. Lang M, Wiesche M, Krcmar H (2018) Criteria for selecting cloud service providers: a Delphi study of quality-of-service attributes. Inf Manage 55(6):746–758
14. Vayghan LA et al (2018) Deploying microservice based applications with Kubernetes: experiments and lessons learned. In: 2018 IEEE 11th international conference on cloud computing (CLOUD). IEEE, pp 970–973
15. Srirama SN, Adhikari M, Paul S (2020) Application deployment using containers with autoscaling for microservices in cloud environment. J Netw Comput Appl 02629
16. Taibi D, Auer F, Lenarduzzi V, Felderer M (2019) From monolithic systems to microservices: an assessment framework. arXiv:1909.08933
17. Kalske M, Mäkitalo N, Mikkonen T (2017) Challenges when moving from monolith to microservice architecture. In: International conference on web engineering. Springer, Cham, pp 32–47

18. Li Z et al (2017) Performance overhead comparison between hypervisor and container based virtualization. In: 2017 IEEE 31st international conference on advanced information networking and applications (AINA). IEEE

19. Salah T et al (2017) Performance comparison between container-based and VM-based services. In: 2017 20th conference on innovations in clouds, internet and networks (ICIN). IEEE, pp 185–190

20. De Benedictis M, Lioy A (2019) Integrity verification of Docker containers for a lightweight cloud environment. Future Gener Comput Syst 97:236–246

21. Bartoletti D, Rymer J, Mines C, Sjoblom S, Turley C (2020) Public cloud development and infrastructure platforms, North America, Q1 2020. The Forrester Wave™. Forrester

22. AWS Homepage. https://aws.amazon.com/lambda/?nc1=h_ls. Last accessed 31 July 2020

23. AWS Homepage. https://aws.amazon.com/ec2/features/?nc1=h_ls. Last accessed 31 July 2020

24. Microsoft Homepage. https://docs.microsoft.com/en-us/azure/virtual-machines/windows/ove rview. Last accessed 01 Aug 2020

25. Google Cloud Homepage. https://cloud.google.com/kubernetes-engine/docs/concepts/kubern etes-engine-overview. Last accessed 02 Aug 2020

26. Google Cloud Homepage. https://cloud.google.com/compute#all-features. Last accessed 02 Aug 2020

Author Index

H. Kim and R. Lee (eds.), *Software Engineering in IoT, Big Data, Cloud and Mobile Computing*, Studies in Computational Intelligence 930,
https://doi.org/10.1007/978-3-030-64773-5

219

Printed in the United States
by Baker & Taylor Publisher Services